Horticulture: Principles and Practices

Horticulture: Principles and Practices

Edited by
Damien Miller

Larsen & Keller
www.larsen-keller.com

Horticulture: Principles and Practices
Edited by Damien Miller
ISBN: 978-1-63549-143-2 (Hardback)

Larsen & Keller
Published by Larsen and Keller Education,
5 Penn Plaza,
19th Floor,
New York, NY 10001, USA

Cataloging-in-Publication Data

Horticulture : principles and practices / edited by Damien Miller.
p. cm.
Includes bibliographical references and index.
ISBN 978-1-63549-143-2
1. Horticulture. 2. Horticulture--Technique. 3. Plant Genetics. 4. Plant diseases. I. Miller, Damien.
SB318 .H67 2017
635--dc23

Printed and bound in China.

For more information regarding Larsen and Keller Education and its products, please visit the publisher's website www.larsen-keller.com

Table of Contents

Preface

This book elucidates new techniques of horticulture and their applications in a multidisciplinary approach. It describes in detail the various concepts and theories related to the field. As a part of agricultural sciences, horticulture deals with the science, art, business and technology of growing plants. The major topics included in the text, which are a sub-part of garden design are landscape restoration, arboriculture, plant conservation, etc. Such selected concepts that redefine the subject have been presented in this text. For all those who are interested in horticulture, this textbook can prove to be an essential guide. It will serve as a source of reference for those interested in the field of horticulture.

To facilitate a deeper understanding of the contents of this book a short introduction of every chapter is written below:

Chapter 1- Horticulture comprehensively studies the methods and technologies utilized in the cultivation and propagation of plants. The plants studied not only include crops but also plants of medicinal value, ornamental plants, shrubs and even trees. This chapter provides a brief overview on the topic of horticulture.

Chapter 2- Horticulture involves nine areas of study, broadly grouped into two sections which are ornamentals and edibles. Arboriculture, floriculture, landscape architecture, olericulture, pomology, viticulture, etc. are some of the themes that are included in this section. The chapter strategically encompasses and incorporates the major components and key concepts of horticulture, providing a complete understanding.

Chapter 3- This chapter explains the varied practices of horticulture like vertical farming, intercropping, controlled environmental agriculture. Techniques of horticulture depend on the region and nature of land and soil under cultivation. The text provides comprehensive information about the principle, practices and techniques in horticulture to all readers.

Chapter 4- Plant genetics is a branch of genetic science that is involved in the mapping and sequencing of the plant genome as well as producing genetically altered plants (GMOs). The chapter delves deep into the predominant methods of modifying plants through the use of the gene gun and the agrobacterium method. A section of the content elaborates on genetically modified trees and crops.

Chapter 5- This chapter discusses the various applications of horticulture in fields like horticultural botany, container gardening, garden designing, urban horticulture etc. These fields use the advancements, technology and techniques of horticulture for various purposes. The content included in this chapter gives the reader in-depth knowledge about the practical applications of horticulture.

Chapter 6- The study of diseases caused by pathogens and harmful ecological conditions is known as plant pathology. This chapter introduces the reader to plant pathology and lists the various pathogens, pests and conditions that lead to infection in plants. The text lists several well-known plant diseases, their symptoms and epidemiology. The topic of disease resistance and the various mechanisms of resistance are tackled in this chapter. Horticulture is best understood in confluence with the major topics listed in the following chapter.

I would like to share the credit of this book with my editorial team who worked tirelessly on this book. I owe the completion of this book to the never-ending support of my family, who supported me throughout the project.

Editor

1

Introduction to Horticulture

Horticulture comprehensively studies the methods and technologies utilized in the cultivation and propagation of plants. The plants studied not only include crops but also plants of medicinal value, ornamental plants, shrubs and even trees. This chapter provides a brief overview on the topic of horticulture.

Horticulture is the branch of agriculture that deals with the art, science, technology, and business of growing plants. It includes the cultivation of medicinal plants, fruits, vegetables, nuts, seeds, herbs, sprouts, mushrooms, algae, flowers, seaweeds and non-food crops such as grass and ornamental trees and plants. It also includes plant conservation, landscape restoration, landscape and garden design, construction, and maintenance, and arboriculture. Inside agriculture, horticulture contrasts with extensive field farming as well as animal husbandry.

A fruit and vegetable market in Milan

Horticulturists apply their knowledge, skills, and technologies used to grow intensively produced plants for human food and non-food uses and for personal or social needs. Their work involves plant propagation and cultivation with the aim of improving plant growth, yields, quality, nutritional value, and resistance to insects, diseases, and environmental stresses. They work as gardeners, growers, therapists, designers, and technical advisors in the food and non-food sectors of horticulture. Horticulture even refers to the growing of plants in a field or garden.

A vertical garden

Horticultural and health-related produce

Etymology

The word *horticulture* is modeled after *agriculture*, and comes from the Greek *χόρτος*, which in Latin became *hortus* "garden" and *cultūra* "cultivation", from *cultus*, the perfect passive participle of the verb *colō* "I cultivate". *Hortus* is cognate with the native English word *yard* (in the meaning of land associated with a building) and also the borrowed word *garden.*

Scope

Horticulture involves nine areas of study, which can be grouped into two broad sections: ornamentals and edibles:

- Arboriculture is the study of, and the selection, plant, care, and removal of, individual trees, shrubs, vines, and other perennial woody plants.
- Turf management includes all aspects of the production and maintenance of turf grass for sports, leisure use or amenity use.
- Floriculture includes the production and marketing of floral crops.
- Landscape horticulture includes the production, marketing and maintenance of landscape plants.

- Olericulture includes the production and marketing of vegetables.
- Pomology includes the production and marketing of pome fruits.
- Viticulture includes the production and marketing of grapes.
- Oenology includes all aspects of wine and winemaking.
- Postharvest physiology involves maintaining the quality of and preventing the spoilage of plants and animals.

Anthropology

Horticulture has a very long history. The study and science of horticulture dates all the way back to the times of Cyrus the Great of ancient Persia, and has been going on ever since, with present-day horticulturists such as Freeman S. Howlett and Luther Burbank. The practice of horticulture can be retraced for many thousands of years. The cultivation of taro and yam in Papua New Guinea dates back to at least 6950–6440 cal BP. The origins of horticulture lie in the transition of human communities from nomadic hunter-gatherers to sedentary or semi-sedentary horticultural communities, cultivating a variety of crops on a small scale around their dwellings or in specialized plots visited occasionally during migrations from one area to the next (such as the "milpa" or maize field of Mesoamerican cultures). In the Pre-Columbian Amazon Rainforest, natives are believed to have used biochar to enhance soil productivity by smoldering plant waste. European settlers called it *Terra Preta de Indio*. In forest areas such horticulture is often carried out in swiddens ("slash and burn" areas). A characteristic of horticultural communities is that useful trees are often to be found planted around communities or specially retained from the natural ecosystem.

Horticulture primarily differs from agriculture in two ways. First, it generally encompasses a smaller scale of cultivation, using small plots of mixed crops rather than large fields of single crops. Secondly, horticultural cultivations generally include a wide variety of crops, even including fruit trees with ground crops. Agricultural cultivations however as a rule focus on one primary crop. In pre-contact North America the semi-sedentary horticultural communities of the Eastern Woodlands (growing maize, squash and sunflower) contrasted markedly with the mobile hunter-gatherer communities of the Plains people. In Central America, Maya horticulture involved augmentation of the forest with useful trees such as papaya, avocado, cacao, ceiba and sapodilla. In the cornfields, multiple crops were grown such as beans (using cornstalks as supports), squash, pumpkins and chilli peppers, in some cultures tended mainly or exclusively by women.

Organizations

The oldest Horticultural society in the world, founded in 1768, is the Ancient Society of York Florists. They still have four shows a year in York, UK.They have a large archive of records, including the original members book dating back to 1768.

Desh Bhagat School of Horticulture Sciences at Desh Bhagat University, Punjab is a prestigious institution in India to encourage the research and promote horticulture techniques throughout the country. The Royal Horticultural Society is a UK charity which oversees several major shows and gardens.

The professional body representing horticulturists in Great Britain and Ireland is the Institute of Horticulture (IOH). Also, the IOH has an international branch for members outside of these islands.

The International Society for Horticultural Science promotes and encourages research and education in all branches of horticultural science.

The American Society of Horticultural Science promotes and encourages research and education in all branches of horticultural science in the Americas.

The Australian Society of Horticultural Science was established in 1990 as a professional society for the promotion and enhancement of Australian horticultural science and industry.

The National Junior Horticultural Association (NJHA) was established in 1934 and was the first organization in the world dedicated solely to youth and horticulture. NJHA programs are designed to help young people obtain a basic understanding of, and develop skills in, the ever expanding art and science of horticulture.

The New Zealand Horticulture Institute.

The Global Horticulture Initiative (GlobalHort) fosters more efficient and effective partnerships and collective action among different stakeholders in horticulture. The organization has a special focus on horticulture for development (H4D), i.e. using horticulture to reduce poverty and improve nutrition worldwide. To be efficient, GlobalHort is organized in a consortium of national and international organizations to collaborate in research, training, and technology-generating activities designed to meet mutually-agreed-upon objectives. GlobalHort is a not-for-profit organization registered in Belgium.

2

Branches of Horticulture

Horticulture involves nine areas of study, broadly grouped into two sections which are ornamentals and edibles. Arboriculture, floriculture, landscape architecture, olericulture, pomology, viticulture, etc. are some of the themes that are included in this section. The chapter strategically encompasses and incorporates the major components and key concepts of horticulture, providing a complete understanding.

Floriculture

Floriculture, or flower farming, is a discipline of horticulture concerned with the cultivation of flowering and ornamental plants for gardens and for floristry, comprising the floral industry. The development, via plant breeding, of new varieties is a major occupation of floriculturists.

A retail greenhouse shows some of the diversity of floricultural plants

Floriculture crops include bedding plants, houseplants, flowering garden and pot plants, cut cultivated greens, and cut flowers. As distinguished from nursery crops, floriculture crops are generally herbaceous. Bedding and garden plants consist of young flowering plants (annuals and perennials) and vegetable plants. They are grown in cell packs (in flats or trays), in pots, or in hanging baskets, usually inside a controlled environment, and sold largely for gardens and landscaping. *Pelargonium* ("geraniums"), *Impatiens* ("busy lizzies"), and *Petunia* are the best-selling bedding plants. The many cultivars of *Chrysanthemum* are the major perennial garden plant in the United States.

Flowering plants are largely sold in pots for indoor use. The major flowering plants are poinsettias, orchids, florist chrysanthemums, and finished florist azaleas. Foliage plants are also sold in pots and hanging baskets for indoor and patio use, including larger specimens for office, hotel, and restaurant interiors.

Cut flowers are usually sold in bunches or as bouquets with cut foliage. The production of cut flowers is specifically known as the cut flower industry. Farming flowers and foliage employs special aspects of floriculture, such as spacing, training and pruning plants for optimal flower harvest; and post-harvest treatment such as chemical treatments, storage, preservation and packaging. In Australia and the United States some species are harvested from the wild for the cut flower market.

Landscape Architecture

Landscape architecture is the design of outdoor public areas, landmarks, and structures to achieve environmental, social-behavioral, or aesthetic outcomes. It involves the systematic investigation of existing social, ecological, and soil conditions and processes in the landscape, and the design of interventions that will produce the desired outcome. The scope of the profession includes landscape design; site planning; stormwater management; environmental restoration; parks and recreation planning; visual resource management; green infrastructure planning and provision; and private estate and residence landscape master planning and design; all at varying scales of design, planning and management. A practitioner in the profession of landscape architecture is called a landscape architect.

Stourhead in Wiltshire, England, designed by Henry Hoare (1705–1785), "the first landscape gardener, who showed in a single work, genius of the highest order"

Definition

Landscape architecture is a multi-disciplinary field, incorporating aspects of botany, horticulture, the fine arts, architecture, industrial design, soil sciences, environmental psychology, geography, and ecology. The activities of a landscape architect can range from the creation of public parks and parkways to site planning for campuses and corporate office parks, from the design of residential estates to the design of civil infrastructure and the management of large wilderness areas or reclamation of degraded landscapes such as mines or landfills. Landscape architects work on structures and external spaces with limitations toward the landscape or park aspect of the design - large or small, urban, suburban and rural, and with "hard" (built) and "soft" (planted) materials, while integrating ecological sustainability. The most valuable contribution can be made at the first stage of a project to generate ideas with technical understanding and creative flair for the design, organization, and use of spaces. The landscape architect can conceive the overall concept and prepare the master plan, from which detailed design drawings and technical specifications are prepared. They can also review proposals to authorize and supervise contracts for the construction work. Other skills include preparing design impact assessments, conducting environmental assessments and audits, and serving as an expert witness at inquiries on land use issues.

Fields of Activity

Royal Botanic Gardens, Kew, London, established 1759
The Palm House built 1844–1848 by Richard Turner to Decimus Burton's designs

Urban design in city squares. Water feature in London, by Tadao Ando who also works with landscapes and gardens

The variety of the professional tasks that landscape architects collaborate on is very broad, but some examples of project types include:

- The planning, form, scale and siting of new developments
- Parks of General design and public infrastructure
- Sustainable development
- Stormwater management including rain gardens, green roofs, groundwater recharge, Green infrastructure, and treatment wetlands

- Landscape design for educational function and site design for public institutions and government facilities
- Parks, botanical gardens, arboretums, greenways, and nature preserves
- Recreation facilities; i.e.: playgrounds, golf courses, theme parks and sports facilities
- Housing areas, industrial parks and commercial developments
- Estate and residence landscape master planning and design
- Highways, transportation structures, bridges, and transit corridors
- Urban design, town and city squares, waterfronts, pedestrian schemes, and parking lots
- Natural park, tourist destination, and recreating historical landscapes, and historic garden appraisal and conservation studies
- Reservoirs, dams, power stations, reclamation of extractive industry applications or major industrial projects and mitigation
- Environmental assessment and landscape assessment, planning advice and land management proposals.
- Coastal and offshore developments and mitigation
- Ecological Design any aspect of design that minimizes environmentally destructive impacts by integrating itself with natural processes and sustainability

Landscape managers use their knowledge of landscape processes to advise on the long-term care and development of the landscape. They often work in forestry, nature conservation and agriculture.

Landscape scientists have specialist skills such as soil science, hydrology, geomorphology or botany that they relate to the practical problems of landscape work. Their projects can range from site surveys to the ecological assessment of broad areas for planning or management purposes. They may also report on the impact of development or the importance of particular species in a given area.

Landscape planners are concerned with landscape planning for the location, scenic, ecological and recreational aspects of urban, rural and coastal land use. Their work is embodied in written statements of policy and strategy, and their remit includes master planning for new developments, landscape evaluations and assessments, and preparing countryside management or policy plans. Some may also apply an additional specialism such as landscape archaeology or law to the process of landscape planning.

Green roof (or more specifically, vegetative roof) designers design extensive and intensive roof gardens for storm water management, evapo-transpirative cooling, sustainable architecture, aesthetics, and habitat creation.

History of Landscape Architecture

For the period before 1800, the history of landscape gardening (later called landscape architecture) is largely that of master planning and garden design for manor houses, palaces and royal properties, religious complexes, and centers of government. An example is the extensive work by André Le Nôtre at Vaux-le-Vicomte for King Louis XIV of France at the Palace of Versailles. The first person to write of *making* a landscape was Joseph Addison in 1712. The term landscape architecture was invented by Gilbert Laing Meason in 1828, and John Claudius Loudon (1783–1843) was instrumental in the adoption of the term landscape architecture by the modern profession. He took up the term from Meason and gave it publicity in his Encyclopedias and in his 1840 book on the *Landscape Gardening and Landscape Architecture of the Late Humphry Repton.*

Orangery at the Palace of Versailles, outside Paris

The practice of landscape architecture spread from the Old to the New World. The term "landscape architect" was used as a professional title by Frederick Law Olmsted in the United States in 1863 and Andrew Jackson Downing (1815–1852), another early American landscape designer, was editor of *The Horticulturist* magazine (1846–52). In 1841 his first book, *A Treatise on the Theory and Practice of Landscape Gardening, Adapted to North America,* was published to a great success; it was the first book of its kind published in the United States. During the latter 19th century, the term landscape architect begun to be used by professional landscapes designers, and was firmly established after Frederick Law Olmsted, Jr. and Beatrix Jones (later Farrand) with others founded the American Society of Landscape Architects (ASLA) in 1899. IFLA was founded at Cambridge, England, in 1948 with Sir Geoffrey Jellicoe as its first president, representing 15 countries from Europe and North America. Later, in 1978, IFLA's Headquarters were established in Versailles.

Relation to Urban Planning

Through the 19th century, urban planning became a focal point and central issue in cities. The combination of the tradition of landscape gardening and the emerging field of urban planning offered Landscape Architecture an opportunity to serve these needs. In the second half of the century, Frederick Law Olmsted completed a series of parks which continue to have a huge influence on the practices of Landscape Architecture today. Among these were Central Park in New York City, Prospect Park in Brooklyn, New York and Boston's Emerald Necklace park system. Jens Jensen designed sophisticated and naturalistic urban and regional parks for Chicago, Illinois, and private estates for the Ford family including Fair Lane and Gaukler Point. One of the original ten founding members of the American Society of Landscape Architects (ASLA), and the only woman, was Beatrix Farrand. She was design consultant for over a dozen universities including: Princeton in Princeton, New Jersey; Yale in New Haven, Connecticut; and the Arnold Arboretum for Harvard in Boston, Massachusetts. Her numerous private estate projects include the landmark Dumbarton Oaks in the Georgetown neighborhood of Washington, D.C.. Since that time, other architects – most notably Ruth Havey and Alden Hopkins—changed certain elements of the Farrand design.

The combination of the traditional landscape gardening and the emerging city planning combined together gave landscape architecture its unique focus.Frederick Law Olmsted used the term 'landscape architecture' using the word as a profession for the first time when designing the Central Park.

Since this period Urban Planning has developed into a separate independent profession that has incorporated important contributions from other fields such as Civil Engineering, Architecture and Public Administration. Urban Planners are qualified to perform tasks independent of landscape architects, and in general, the curriculum of landscape architecture programs do not prepare students to become urban planners.

Landscape architecture continues to develop as a design discipline, and to respond to the various movements in architecture and design throughout the 20th and 21st centuries. Thomas Church was a mid-century landscape architect significant in the profession. Roberto Burle Marx in Brazil combined the International style and native Brazilian plants and culture for a new aesthetic. Innovation continues today solving challenging problems with contemporary design solutions for master planning, landscapes, and gardens.

Ian McHarg was known for introducing environmental concerns in landscape architecture. He popularized a system of analyzing the layers of a site in order to compile a complete understanding of the qualitative attributes of a place. This system became the foundation of today's Geographic Information Systems (GIS). McHarg would give every qualitative aspect of the site a layer, such as the history, hydrology, topography, vegetation, etc. GIS software is ubiquitously used in the landscape architecture profession today to analyze materials in and on the Earth's surface and is similarly used by Urban Planners, Geographers, Forestry and Natural Resources professionals, etc.

Profession

In many countries, a professional institute, comprising members of the professional community, exists in order to protect the standing of the profession and promote its interests, and sometimes also regulate the practice of landscape architecture. The standard and strength of legal regulations governing landscape architecture practice varies from nation to nation, with some requiring licensure in order to practice; and some having little or no regulation. In North America, Europe, Australia and New Zealand, landscape architecture is a regulated profession.

Australia

The Australian Institute of Landscape Architects (AILA) provides non statutory professional recognition for landscape architects. Once recognized by AILA, landscape architects use the title 'Registered Landscape Architect'. Across the eight states and territories within Australia, there is a mix of requirements for landscape architects to be 'Registered', however it is not always a statutory requirement to be registered with AILA to practice use the term "Landscape Architect".

Any regulations or requirements are state based, not national. The AILA's system of professional recognition is a national system overseen by AILA's National Office in Canberra. Non (A.I.L.A) Landscape Architects are professionals who are also paid to undertake a specialised set of tasks and to complete them for a fee.

Some agencies require AILA professional recognition or registration as part of the pre-requisite for contracts. Landscape architects within Australia find that some contracts and competitions require the AILA recognition or 'registration' as the basis of demonstrating a professional status. To apply for AILA Registration, an applicant usually needs to satisfy a number of pre-requisites, including university qualification, two years of practice and a record of continuing professional practice. The application is in two stages: (1) A minimum 12 months of mentoring and assessment; and (2) oral assessment/interview. Professional recognition includes a commitment to continue professional development. A.I.L.A Registered Landscape Architects are required to report annually on their continuing professional development.

Landscape Architecture within Australia covers a broad spectrum of design, advice and research. From specialist design services for commercial and government developments through to professional advice as an expert witness, the range of tasks delivered by Australian Landscape Architect's is diverse and interesting.

The harsh Australian environment also provide numerous challenges that must be overcome. Australia is the driest inhabited continent in the world and this characteristic determines particular requirements through design including specific species selection and careful consideration of natural resources such as rainfall and topography.

Canada

In Canada, landscape architecture, like law and medicine, is a self-regulating profession pursuant to provincial statute. For example, Ontario's profession is governed by the Ontario Association of Landscape Architects pursuant to the *Ontario Association of Landscape Architects Act*. Landscape architects in Ontario, British Columbia, and Alberta must complete the specified components of

L.A.R.E (Landscape Architecture Registration Examination) as a prerequisite to full professional standing.

Provincial regulatory bodies are members of a national organization, the Canadian Society of Landscape Architects / L'Association des Architectes Paysagistes du Canada (CSLA-AAPC), and individual membership in the CSLA-AAPC is obtained through joining one of the provincial or territorial components.

Italy

AIAPP (Italian Association of Landscape Architecture) is the Italian association of professional landscape architects formed in 1950 and is a member of IFLA and IFLA Europe (formerly known as EFLA). AIAPP is in the process of contesting this new law which has given the Architects' Association the new title of Architects, Landscape Architects, Planners and Conservationists whether or not they have had any training or experience in any of these fields other than Architecture. In Italy, there are several different professions involved in landscape architecture:

- Architects
- Landscape designers
- Doctor landscape agronomists and Doctor landscape foresters, often called Landscape agronomists.
- Agrarian Experts and Graduated Agrarian experts.

New Zealand

The New Zealand Institute of Landscape Architects (NZILA) is the professional body for Landscape Architects in NZ www.nzila.co.nz.

In April 2013, NZILA jointly with AILA, hosted the 50th International Federation of Landscape Architects (IFLA) World Congress in Auckland, New Zealand. The World Congress is an international conference where Landscape Architects from all around the globe meet to share ideas around a particular topic.

Within NZ, Members of NZILA when they achieve their professional standing, can use the title Registered Landscape Architect NZILA.

NZILA provides an education policy and an accreditation process to review education programme providers; currently there are three accredited undergraduate Landscape Architecture programmes in New Zealand. Lincoln University also has an accredited masters programme in landscape architecture.

Republic of Ireland

The professional body in Ireland for landscape architects is the Irish Landscape Institute (ILI) www.irishlandscapeinstitute.com. The ILI is an affiliate body to the European Federation for Landscape Architecture (EFLA) and IFLA. The ILI was formed in 1993 to merge the disciplines of landscape architecture and landscape horticulture. It continues to promote the profession by its accreditation of the degree programme in Dublin, certification of Continuing Professional Development (CPD) for landscape architects, administration of professional practice examinations, advice on development of policy at national level and organisation of conferences, lectures and design awards. The ILI is a member institute of the Urban Forum, representing professional bodies involved in urban spatial disciplines of engineering, architecture, planning, quantity surveying and landscape architecture.

The profession has gained in status and numbers due to the construction boom of the past decade and raising of standards of Irish design. There is still no registration of title in Ireland and the profession is unregulated, but there is increasing awareness of the profession and of status of the ILI. Landscape architects in Ireland work in private practice, public sector bodies at local government level and in some bodies such transport and national heritage and in the academic sector. The demand for landscape architects is often associated with strategic infrastructure projects due to Ireland's recent major infrastructural investments. Landscape architects are employed in design of: green infrastructure, public realm, institutional/medical/industrial campuses and settings, parks, play facilities, transport (road/rail/cycle/port) corridors, retail complexes, residential estates (including plans for remediation of now-abandoned housing 'ghost' estates), village improvements, accessibility audits, graveyard restoration schemes, wind farms, wetland drainage systems and coastal zones. They are also significantly employed in preparation/review of statutory impact assessment reports on landscape, visual and ecological impacts of design proposals.

South Africa

In May 1962, Joane Pim, Ann Sutton, Peter Leutscher and Roelf Botha (considered the forefathers of the profession in South Africa) established the Institute for Landscape Architects, now known as the Institute for Landscape Architecture in South Africa (ILASA). ILASA is a voluntary organisation registered with the South African Council for the Landscape Architectural Profession. It consists of three regional bodies namely, Gauteng, KwaZula-Natal and the Western Cape. ILASA's mission is to advance the profession of landscape architecture and uphold high standards of professional service to its members, and to represent the profession of landscape architecture in any matter which may affect the interests of the members of the Institute. ILASA holds the country's membership with The International Federation of Landscape Architects (IFLA).

In South Africa, the profession is regulated by The South African Council for the Landscape Architectural Profession (SACLAP), established as a statutory council in terms of Section 2 of the South African Council for the Landscape Architectural Profession Act – Act 45 of 2000. The Council evolved out of the Board of Control for Landscape Architects (BOCLASA), which functioned under the Council of Architects in terms of The Architectural Act, Act 73 of 1970. SACLAP's mission is to establish, direct, sustain and ensure a high level of professional responsibilities and ethical conduct within the art and science of landscape architecture with honesty, dignity and integrity in the broad interest of public health, safety and welfare of the community.

After completion of an accreditted under-graduate and/or post-graduate qualification in landscape architecture at either the University of Cape Town or the University of Pretoria, or landscape technology at the Cape Peninsula University of Technology, professional registration is attained via a mandatory mentored candidacy period (minimum of two years) and sitting of the professional registration exam. After successfully completing the exam, the individual is entitled to the status of Professional Landscape Architect or Professional Landscape Technologist.

United Kingdom

The UK's professional body is the Landscape Institute (LI). It is a chartered body which accredits landscape professionals and university courses. At present there are fifteen accredited programmes in the UK. Membership of the LI is available to students, academics and professionals, and there are over 3,000 professionally qualified members.

The Institute provides services to assist members including support and promotion of the work of landscape architects; information and guidance to the public and industry about the specific expertise offered by those in the profession; and training and educational advice to students and professionals looking to build upon their experience.

In 2008, the LI launched a major recruitment drive entitled "I want to be a Landscape Architect" to encourage the study of Landscape Architecture. The campaign aims to raise the profile of landscape architecture and highlight its valuable role in building sustainable communities and fighting climate change.

United States

In the United States, Landscape Architecture is regulated by individual state governments. For a landscape architect, obtaining licensure requires advanced education and work experience, plus passage of the national examination. Several states require passage of a state exam as well. In the United States licensing is overseen both at the state level, and nationally by the Council of Landscape Architectural Registration Boards (CLARB). Landscape architecture has been identified as an above-average growth profession by the US Bureau of Labor Statistics and was listed in *U.S. News & World Report*'s list of Best Jobs to Have in 2006, 2007, 2008, 2009 and 2010. The national trade association for United States landscape architects is the American Society of Landscape Architects.

Examples of Landscape Architecture

Monastery garden Bebenhausen Abbey in Baden-Württemberg, Germany

Woodland in Saihō-ji, Japan

Olericulture

Olericulture is the science of vegetable growing, dealing with the culture of non-woody (herbaceous) plants for food.

Olericulture is the production of plants for use of the edible parts. Vegetable crops can be classified into 9 major categories:

- Potherbs and greens – spinach and collards
- Salad crops – lettuce, celery
- Cole crops – cabbage and cauliflower
- Root crops (tubers) – potatoes, beets, carrots, radishes
- Bulb crops – onions, leeks
- Legumes – beans, peas
- Cucurbits – melons, squash, cucumber
- Solanaceous crops – tomatoes, peppers
- Sweet corn

Olericulture deals with the production, storage, processing and marketing of vegetables. It encompasses crop establishment, including cultivar selection, seedbed preparation and establishment of vegetable crops by seed and transplants.

It also includes maintenance and care of vegetable crops as well commercial and non-traditional vegetable crop production including organic gardening and organic farming; sustainable agriculture and horticulture; hydroponics; and biotechnology.

Pomology

Pomology (from Latin *pomum* (fruit) + *-logy*) is a branch of botany that studies and cultivates fruit. The denomination fruticulture—introduced from Romance languages (from Latin *fructus* and *cultura*)—is also used.

Illustration of the 'Willermoz' pear by Alexandre Bivort from *Album de Pomologie* (1848–1852)

Pomological research is mainly focused on the development, cultivation and physiological studies of fruit trees. The goals of fruit tree improvement include enhancement of fruit quality, regulation of production periods, and reduction of production cost. One involved in the science of pomology is called a pomologist.

History

Pomology has been an important area of research for centuries.

United States

During the mid-19th century in the United States, farmers were expanding fruit orchard programs in response to growing markets. At the same time, horticulturists from the USDA and agricultural colleges were bringing new varieties to the United States from foreign expeditions, and developing experimental lots for these fruits. In response to this increased interest and activity, USDA established the Division of Pomology in 1886 and named Henry E. Van Deman as chief pomologist. An important focus of the division was to publish illustrated accounts of new varieties and to disseminate research findings to fruit growers and breeders through special publications and annual reports. During this period Andrew Jackson Downing and his brother Charles were prominent in Pomology and Horticulture, producing *The Fruits and Fruit Trees of America* (1845).

The introduction of new varieties required exact depiction of the fruit so that plant breeders could accurately document and disseminate their rearch results. Since the use of scientific photography was not widespread in the late 19th Century, USDA commissioned artists to create watercolor illustrations of newly introduced cultivars. Many of the watercolors were used for lithographic reproductions in USDA publications, such as the *Report of the Pomologist* and the *Yearbook of Agriculture*.

Today, the collection of approximately 7,700 watercolors is preserved in the National Agricultural Library's Special Collections, where it serves as a major historic and botanic resource to a variety of researchers, including horticulturists, historians, artists, and publishers..

The study of pomology has somewhat dwindled over the past century..

Viticulture

Viticulture (from the Latin word for *vine*) is the science, production, and study of grapes. It deals with the series of events that occur in the vineyard. It is a branch of the science of horticulture.

Wine grapes

Vineyard in Brhlovce, Slovakia.

While the native territory of *Vitis vinifera*, the common grape vine, ranges from Western Europe to the Persian shores of the Caspian Sea, the vine has demonstrated high levels of adaptability to new environments. For this reason, viticulture can be found on every continent except Antarctica.

Duties of the viticulturist include monitoring and controlling pests and diseases, fertilizing, irrigation, canopy management, monitoring fruit development and characteristics, deciding when to harvest, and vine pruning during the winter months. Viticulturists are often intimately involved with winemakers, because vineyard management and the resulting grape characteristics provide the basis from which winemaking can begin. A great number of varieties are now approved in the European Union as true grapes for wine-growing and viticulture, because of better characteristics for diseases.

History

The earliest evidence of grape vine cultivation and winemaking dates back 7,000 years. The history of viticulture is closely related to the history of wine, with evidence that humans cultivated wild grapes to make wine as far back as the Neolithic period. Evidence suggests that some of the earliest domestication of *Vitis vinifera* occurred in the area of the modern countries Georgia and Armenia. The oldest-known winery was discovered in the "Areni-1" cave in Vayots Dzor, Armenia. Dated to c. 4100 BC, the site contained a wine press, fermentation vats, jars, and cups. Archaeologists also found *V. vinifera* seeds and vines. Commenting on the importance of the find, McGovern said, "The fact that winemaking was already so well developed in 4000 BC suggests that the technology probably goes back much earlier." There is also evidence of grape domestication in the Near East in the early Bronze Age, around 3200 BC.

Evidence of ancient viticulture is provided by cuneiform sources (ancient writing on clay tablets), plant remains, historical geography, and archaeological excavations. The remnants of ancient wine jars have been used to determine the culture of wine consumption and cultivated grape species. In addition to winemaking, grapes have been grown for the production of raisins.

The earliest act of cultivation appears to have been the favoring of hermaphroditic members of the *Vitis vinifera* species over the barren male vines and the female vines, which were dependent on a nearby male for pollination. With the ability to pollinate itself, over time the hermaphroditic vines were able to sire offspring that were consistently hermaphroditic.

At the end of the 5th century BC, the Greek historian Thucydides wrote:

> "The people of the Mediterranean began to emerge from barbarism when they learnt to cultivate the olive and the vine."

The period that Thucydides was most likely referencing was the time between 3000 BC and 2000 BC, when viticulture emerged in force in Asia Minor, Greece, and the Cyclades Islands of the Aegean Sea. During this period, grape cultivation developed from an aspect of local consumption to an important component of international economies and trade.

Roman

From 1200 BC to 900 BC, the Phoenicians developed viticulture practices that were later used in Carthage. Around 500 BC, the Carthaginian writer Mago recorded such practices in a two-volume work that was one of the few artifacts to survive the Roman destruction of Carthage during the Third Punic War. The Roman statesman Cato the Elder was influenced by these texts, and around 160 BC he wrote *De Agricultura,* which expounded on Roman viticulture and agriculture. Around 65 AD, the Roman writer Columella produced the most detailed work on Roman viticulture in his twelve-volume text *De Re Rustica*. Columella's work is one of the earliest to detail trellis systems for raising vines off the ground. Columella advocated the use of stakes versus the previously accepted practice of training vines to grow up along tree trunks. The benefits of using stakes over trees was largely to minimize the dangers associated with climbing trees, which was necessary to prune the dense foliage in order to give the vines sunlight, and later to harvest them.

Roman expansion across Western Europe brought Roman viticulture to the areas that would become some of the world's best-known wine-growing regions: the Spanish Rioja, the German Mosel, and the French Bordeaux, Burgundy and Rhône. Roman viticulturists were among the first to identify steep hillsides as one of the better locations to plant vines, because cool air runs downhill and gathers at the bottom of valleys. While some cool air is beneficial, too much can rob the vine of the heat it needs for photosynthesis, and in winter it increases the risk of frost.

Medieval

In the Middle Ages, Catholic monks (particularly the Cistercians) were the most prominent viticulturists of the time period. Around this time, an early system of *Metayage* emerged in France with laborers (*Prendeur*) working the vineyards under contractual agreements with the landowners (*Bailleur*). In most cases, the *prendeurs* were given flexibility in selecting their crop and developing their own vineyard practice. In northern Europe, the weather and climate posed difficulties for grape cultivation, so certain species were selected that better suited the environment. Most vineyards grew white varieties of grape, which are more resistant to the damp and cold climates. A couple species of red grape, such as the Pinot Noir, were also introduced.

Les Très Riches Heures du duc de Berry dates back to 1416 and depicts horticulture and viticulture in France. The images illustrate peasants bending down to prune grapes from vines behind castle walls. Additional illustrations depict grape vines being harvested, with each vine being cut to three spurs around knee height.

Many of the viticultural practices developed in this time period would become staples of European viticulture till the 18th century. Varietals were studied more intently to see which vines were the most suitable for a particular area. Around this time, an early concept of *terroir* emerged as wines from particular places began to develop a reputation for uniqueness. The concept of pruning for quality over quantity emerged, mainly through Cistercian labors, though it would create conflict between the rich landowners who wanted higher quality wines and the peasant laborers whose livelihood depended on the quantity of wine they could sell. The Riesling is the famous example for higher quality of wine. In 1435 Count John IV. of Katzenelnbogen started this successful tradition.

In Burgundy, the Cistercian monks developed the concept of *cru* vineyards as homogeneous pieces of land that consistently produce wines each vintage that are similar. In areas like the Côte-d'Or the monks divided the land into separate vineyards, many of which are still around today—like Montrachet and La Romanée.

In Mythology

In Greek mythology, the demigod Dionysus (Roman name: Bacchus) – son of Zeus – invented the grapevine and the winepress. When his closest satyr friend died trying to bring him the vine that Dionysus thought was important, Dionysus forced the vine to bear fruit and he later invented wine. His fame spread and he finally became a God.

The Grape Vine

The vast majority of the world's wine producing regions are found between the temperate latitudes of 30° and 50° in each hemisphere. Within these bands, the annual mean temperatures are between 10 and 20 °C (50 and 68 °F). The presence of large bodies of water and mountain ranges can have positive effects on the climate and vines. Nearby lakes and rivers can serve as protection for drastic temperature drops at night by releasing the heat that the water has stored during the day to warm the vines.

The Berry

The grape is classified as a berry. On the vine, grapes are organized through systems known as clusters. Grape clusters can vary in compactness which can result in long clusters (resulting in the grapes spreading out) or short clusters (resulting in grapes packed together). In some grape species, clusters ripen collectively which allow them to be harvested together. For others, grapes may ripen individually within a cluster. Each grape berry contains a pedicel which attaches to the rachis. The main function of the rachis is to allow the grapes receive their water and nutrients. The pollination and fertilization of grapes results in one to four seeds within the berry. When fertilization does not occur, seedless grapes are formed, which are sought after for the production of raisins. Regardless of pollination and fertilization, most plants will produce around 100 to 200 grapes.

The skin of the grape accounts for 5 to 20% of the total weight of a grape depending on the variety. When grape skin ripens, it contains the majority of the aromatic substances and tannin. These factors become important in winemaking for methods including color extraction or aroma dissolution.

Although the skin contains the majority of the tannin, small percentages can be found throughout the grape and during all of its developmental stages. However, the tannin's most important role is during the grape's ripening stage as its function is to formulate color and body shape.

Growing Vines

Although many factors can affect the overall quality of a grape vine, the three most important are climate, slope, and soil.

Climate

Climate is the most significant factor in determining a grape's inherent qualities. Each grape species has a uniquely preferred environment for ideal growing. Because climates vary from country to country, selecting the best strain is an important decision in grape cultivation. In addition, because climatic factors such as temperature and rain can be unpredictable and uncontrollable, each year will produce unique qualities and yields of grapes.

Grape vines need approximately 1300–1500 hours of sunshine during the growing season and around 690 millimetres (27 in) of rainfall throughout the year in order to produce grapes suitable for winemaking. In ideal circumstances, the vine will receive most of the rainfall during the winter and spring months: rain at harvesttime can create many hazards, such as fungal diseases and berry splitting. The optimum weather during the growing season is a long, warm summer that allows the grapes the opportunity to ripen fully and to develop a balance between the levels of acids and sugars in the grape. Hot and sunny climates have a frost-free growing season of 200 days or more. These climates allow grapes to ripen faster with higher sugar levels and lower acidity. Cooler climates have a frost-free growing season of around 150–160 days. Cooler seasons force the grapes to ripen earlier which produces a fresher and more acidic harvest. In general, the average yearly temperature for most crops should average around 15 °C (59 °F) in order for the highest quality to be achieved in each grape.

Summer: Ideal temperatures in summer average around 22 °C (72 °F). Ideal summer temperatures enable fruits to ripen. Temperature and sunshine are the most important factors in ripening.

Winter: Ideal temperatures in winter average around 3 °C (37 °F). Ideal winter temperatures are necessary to allow grape vines to enter their resting phase. If temperatures fall too low, the crops can be injured.

Spring and Fall: Spring and fall are critical seasons for grape development, because the plants are susceptible to frost damage, which can injure the fruiting buds. Wet weather in spring can increase the odds of mildew formation. To prevent mildew, some farms introduce devices such as heaters or large fans in vineyards. However, such solutions can be costly.

Slope

Hillsides and slopes are preferred over flatter terrain: vines growing on a slope can receive a greater intensity of the sun's rays, with sunshine falling on an angle perpendicular to the hillside. In flatter terrain, the intensity of the sunlight is diluted as it spreads out across a wider surface area. Small slopes that are elevated above surrounding ground are the best and safest places for crops, because these small elevations are less prone to frost. Additionally, a slope affords better drainage, obviating the possibility that the vine might sit in overly moist soil. In cooler regions of the northern hemisphere, south-facing slopes receive more hours of sunlight and are preferred; in warmer climes, north-facing slopes are preferred. In the southern hemisphere, these orientations are reversed.

Soil

Quality soil is important to allow plants to have better root systems. The growth and health of a vine can be affected if the soil quality is poor. Different grape species prefer various soil conditions, although there are general quality factors. Favorable soil conditions include: aeration, loose texture, good drainage and moderate fertility. Drainage factors are cited as the most important soil characteristic to affect grape vine growth. When root growth is restricted due to bad soil, vine growth and fruit yields lessen and plant survival rates can dip to only a few years.

Hazards

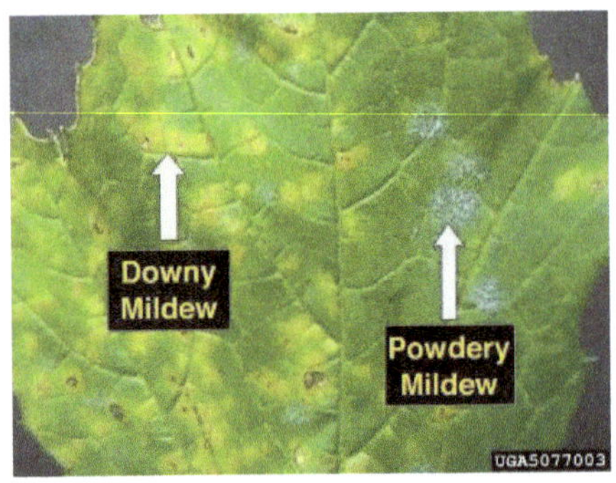

Examples of downy and powdery mildew on a grape leaf.

Manual grape gathering

There are many hazards that a viticulturist needs to be aware of when growing vines. These hazards can have an adverse effect on the wine produced from the grape or kill the vine itself. When the vine is flowering it is very susceptible to weather hazards such as strong winds and hail. Cold temperatures during this period can also bring the onset of *millerandage* which produces clusters with no seeds and varying sizes. Too much heat can have the opposite reaction and produce Coulure that causes grape clusters to either drop to the ground or not fully develop.

Viticultural hazards include:

- Oidium: A powdery mildew which can attack all green parts of the vine. If left untreated, odium can be terminal for the plant. It thrives in cooler temperatures and in the shade. Some North American vine species have evolved to show resistance to the mildew (3:37).
- Downy mildew: Also known as peronospora, thrives in high temperatures and humidity and produces stains on leaves. Peronospora can be treated by spraying plants with copper sulphate. Most American vines are resistant, excluding *Vitis vinifera.*
- Fanleaf: can lead to deformity, yellowing of the leaves, and smaller crop yields. Fanleaf is a virus spread by nematodes which breed in the vine stem. There is no cure for the plant, and the best action to take is to remove the infected plants and allow remaining roots in the soil to rot.
- Frost
- Phylloxera
- Plant virus

Green Harvest

A green harvest is the removal of immature grape bunches, typically for the purpose of increasing yield. The removal of the bunches while they are still green induces the vine to put all its energy into developing the remaining grapes. In theory this results in better ripening and the development of more numerous and mature flavour compounds. In the absence of a green harvest, a healthy, vigorous vine can produce dilute, unripe grapes.

In Europe, many appellations restrict the yield permitted from a given area, so there is even more incentive to perform green harvesting when presented with excess crop. Often the excess must be sold for a pittance and used for industrial alcohol production rather than wine.

While the concept of thinning or sacrificing part of the grape crop, i.e. green harvesting, with the aim of improving the quality of the remaining grapes, predates modern critics, the practice has increased in recent times in vineyards found in California and areas where the grapes grow easily. (McCoy)

Field Blend

A field blend is a wine that is produced from two or more different grape varieties interplanted in the same vineyard. In the days before precise varietal identification, let alone rigorous clonal selection, a vineyard might be planted by taking cuttings from another vineyard and therefore approximately copying its genetic makeup. This meant that one vine could be Zinfandel and the next Carignan. When making wine with little equipment to spare for separate vinification of different varieties, field blends allowed effortless, though inflexible, blending.

Mechanical harvesting of Sauvignon blanc grapes in Côtes de Duras France

Fermentation tanks are now cheap enough that the field blend is an anachronism, and almost all wines are assembled by blending from smaller, individual lots. However, in California some of the oldest (and lowest-yielding) Zinfandel comes from vineyards that are field-blended. Ridge Vineyards owns the Lytton Springs vineyards in Sonoma County, which were planted from 1900 to 1905 in what Ridge calls "a traditional field blend of about seventy percent Zinfandel, twenty percent Petite Sirah, and ten percent Grenache and Carignan."

Gemischter Satz (*Mixed set*) is a wine term in German equivalent to a field blend, which means that grapes of different varieties are planted, harvested and vinified together. In older times, this was common, but the practice has almost stopped. It is, however, a specialty of Vienna.

Arboriculture

An arborist practicing arboriculture: using a chainsaw to fell a eucalyptus tree in a park at Kallista, Victoria.

Arboriculture is the cultivation, management, and study of individual trees, shrubs, vines, and other perennial woody plants. The science of arboriculture studies how these plants grow and respond to cultural practices and to their environment. The practice of arbori-culture includes cultural techniques such as selection, planting, training, fertilization, pest and pathogen control, pruning, shaping, and removal.

A person who practices or studies arboriculture can be termed an 'arborist' or an 'arboriculturist'. A 'tree surgeon' is more typically someone who is trained in the physical maintenance and manipulation of trees and therefore more a part of the arboriculture process rather than an arborist. Risk management, legal issues, and aesthetic considerations have come to play prominent roles in the practice of arboriculture. Businesses often need to hire arboriculturists to complete "tree hazard surveys" and generally manage the trees on-site to fulfill occupational safety and health obligations.

Arboriculture is primarily focused on individual woody plants and trees maintained for permanent landscape and amenity purposes, usually in gardens, parks or other populated settings, by arborists, for the enjoyment, protection, and benefit of human beings. It falls under the general umbrella of horticulture.

Tropical Horticulture

Tropical horticulture is a branch of horticulture that studies and cultivates plants in the tropics, i.e., the equatorial regions of the world. The field is sometimes known by the portmanteau "TropHort".

Latex being collected from an incised rubber tree (*Hevea brasiliensis*), and a bucket of collected latex

Tropical horticulture includes plants such as perennial woody plants (arboriculture), ornamentals (floriculture), vegetables (olericulture), and fruits (pomology) including grapes (viticulture). The origin of many of these crops is not in the tropics but in temperate zones. Their adoption to tropical climatic conditions is an objective of breeding. Many important crops, however, are indigenous to the tropics. The latter embrace perennial crops such as oil palm, vegetables including okra, field crops such as rice and sugarcane, and particularly fruits including pineapple, banana, papaya, and mango.

Since the tropics represent 36 percent of the earth's surface and 20 percent of its land surface, the potential of tropical horticulture is huge. In contrast to temperate regions, environmental conditions in the tropics are defined less by seasonal temperature fluctuations and more by seasonality of precipitation. Thus the climate in the greater part of the tropics is characterized by distinct wet and dry seasons, although such variation is reduced in locations closer to the equator (±5° latitude). Temperature conditions in the tropics are affected by elevation, in which contrasting warmer and colder climate areas in the tropics can be differentiated, and highland areas in the tropics can consequently be more favourable for production of temperate plant species than are lowland areas.

Types of Plants

Both vascular and non-vascular plants grow in tropical environments. Plants indigenous to the tropics are usually cold sensitive and adapted to receiving high levels of solar radiation. They are sensitive to small variations in photoperiod ("short day" plants), and can be adapted to extended drought, high precipitation and/or distinct wet and dry seasons. High night temperatures are a major hindrance to adopting temperate crops (e.g., tomatoes) to the tropical lowlands. Furthermore, such conditions promote high respiration rates of plants, resulting in comparably lower net photosynthesis rates.

References

- Sir Geoffrey Jellicoe, Susan Jellicoe, The Landscape of Man: Shaping the Environment from Prehistory to the Present Day ISBN 9780500274316
- Fraga, H., Garcia de C. A. I., Malheiro, A.C., Santos, J.A., 2016. Modelling climate change impacts on viticultural yield, phenology and stress conditions in Europe. Global Change Biology: doi:10.1111/gcb.13382.
- Corbett, John. "Ian McHarg: Overlay Maps and the Evaluation of Social and Environmental Costs of Land Use Change". Center for Spartially Integrated Social Science. Retrieved 4 June 2014.
- Van Assche, K., Beunen, R., Duineveld, M., & de Jong, H. (2013). Co-evolutions of planning and design: Risks and benefits of design perspectives in planning systems. Planning Theory, 12(2), 177-198.
- "Katzenelnbogener Weltrekorde: Erster RIESLING und erste BRATWURST!". Graf-von-katzenelnbogen.de. Retrieved 2012-02-12.
- "Floriculture News" (PDF). No. 64. The Department of Agriculture, Western Australia. May 2005. Retrieved September 17, 2012.
- David Keys (28 December 2003). "Now that's what you call a real vintage: professor unearths 8,000-year-old wine". The Independent. independent.co.uk. Retrieved 13 January 2011.
- Mark Berkowitz (September–October 1996). "World's Earliest Wine". Archaeology. Archaeological Institute of America. 49 (5). Retrieved 13 January 2011.
- Thomas H. Maugh II (11 January 2011). "Ancient winery found in Armenia". Los Angeles Times. Los Angeles Times Media Group. Retrieved 13 January 2011.

3

Principles, Practices and Techniques in Horticulture

This chapter explains the varied practices of horticulture like vertical farming, intercropping, controlled environmental agriculture. Techniques of horticulture depend on the region and nature of land and soil under cultivation. The text provides comprehensive information about the principle, practices and techniques in horticulture to all readers.

Vertical Farming

Lettuce grown in indoor vertical farming system.

Vertical farming is the practice of producing food in vertically stacked layers, vertically inclined surfaces and/or integrated in other structures. The modern idea of vertical farming uses controlled-environment agriculture (CEA) technology, where all environmental factors can be controlled. These facilities utilize artificial control of light, environmental control (humidity, temperature, gases...) and fertigation. Some vertical farms use techniques similar to greenhouses, where natural sunlight can be augmented with artificial lighting and metal reflectors.

Types

The term "vertical farming" was coined by Gilbert Ellis Bailey in 1915 in his book *Vertical Farming*. His use of the term differs from the current meaning—he wrote about farming with a special interest in soil origin, its nutrient content and the view of plant life as "vertical" life forms, specifically relating to root structures underground. Modern usage refers to skyscrapers using some degree of natural light.

Mixed-use Skyscrapers

Mixed-use skyscrapers were proposed and built by architect Ken Yeang. Yeang proposes that instead of hermetically sealed mass-produced agriculture that plant life should be cultivated within open air, mixed-use skyscrapers for climate control and consumption (i.e., a personal or communal planting space as per the needs of the individual). This version of vertical farming is based upon personal or community use rather than the wholesale production and distribution plant life that aspires to feed an entire city. It thus requires less of an initial investment than Despommier's "vertical farm". However, neither Despommier nor Yeang are the conceptual originators, nor is Yeang the inventor of vertical farming in skyscrapers.

Plantagon is building the world's first industrial-scale multifunctional vertical farm in Linköping, Sweden.

Despommier's Skyscrapers

Dickson Despommier shares his ideas about how "vertical farming" can help reduce hunger by changing the way we use land for agriculture. Photography by Kris Krüg

Ecologist Dickson Despommier argues that vertical farming is legitimate for environmental reasons. He claims that the cultivation of plant life within skyscrapers will require less embedded energy and produce less pollution than some methods of producing plant life on natural landscapes. He moreover claims that natural landscapes are too toxic for natural, agricultural production, despite the ecological and environmental costs of extracting materials to build skyscrapers for the simple purpose of agricultural production.

Vertical farming according to Despommier thus discounts the value of natural landscape in exchange for the idea of "skyscraper as spaceship." Plant life is mass-produced within hermetically sealed, artificial environments that have little to do with the outside world. In this sense, they could be built anywhere regardless of the context. Although climate control, lighting, and other costs of maintenance have been posited as potentially stifling to bringing this concept to fruition, advocates have countered that an important feature of future vertical farms will be the integration of renewable energy technology, be it solar panels, wind turbines, water capture systems, and probably some combination of the three. The vertical farm is designed to be sustainable, and to enable nearby inhabitants to work at the farm.

Despommier's concept of the vertical farm emerged in 1999 at Columbia University. It promotes the mass cultivation of plant life for commercial purposes in skyscrapers.

Stackable Shipping Containers

Several companies have brought forth the concept of stacking recycled shipping containers in urban settings. Freight Farms produces a "leafty green machine" that is a complete farm-to-table system outfitted with vertical hydroponics, LED lighting and intuitive climate controls built within a 12 m × 2.4 m shipping container. Podponics has built a large scale vertical farm in Atlanta consisting of over 100 stacked "growpods". A similar farm is currently under construction in Oman.

History

A commercial high-rise farm such as the 'Vertical Farm' has never been built, yet extensive photographic documentation and several historical books on the subject suggest that research on the subject was not diligently pursued. New sources indicate that a tower hydroponicum existed in Armenia prior to 1951.

Proponents argue that, by allowing traditional outdoor farms to revert to a natural state and reducing the energy costs needed to transport foods to consumers, vertical farms could significantly alleviate climate change produced by excess atmospheric carbon. Critics have noted that the costs of the additional energy needed for artificial lighting, heating and other vertical farming operations would outweigh the benefit of the building's close proximity to the areas of consumption. However, a recent study published in the Journal of Agricultural Engineering and Biotechnology has utilized inexpensive metal reflectors to supply sunlight to the plants.

One of the earliest drawings of a tall building that cultivates food was published in Life Magazine in 1909. The reproduced drawings feature vertically stacked homesteads set amidst a farming landscape. This proposal can be seen in Rem Koolhaas's Delirious New York. Koolhaas wrote that this 1909 theorem is 'The Skyscraper as Utopian device for the production of unlimited numbers of virgin sites on a metropolitan location' (1994, 82).

Other architectural proposals that provide the seeds for the Vertical Farm project include Le Corbusier's Immeubles-Villas (1922) and SITE's Highrise of homes (1972). SITE's Highrise of homes, is a near revival of the 1909 Life Magazine Theorem. In fact, built examples of tower hydroponicums are quite well documented in the canonical text of "The Glass House" by John Hix. Images of the vertical farms at the School of Gardeners in Langenlois, Austria, and the glass tower at the Vienna International Horticulture Exhibition (1964) clearly show that vertical farms existed more than 40 years prior to contemporary discourse on the subject. Although architectural precedents remain valuable, the technological precedents that make vertical farming possible can be traced back to horticultural history through the development of greenhouse and hydroponic technology. Early building types or Hydroponicums were developed, integrating hydroponic technology into building systems. These horticultural building systems evolved from greenhouse technology, and paved the way for the modern concept of the vertical farm. The British Interplanetary Society developed a hydroponicum for lunar conditions and other building prototypes were developed during the early days of space exploration. During this era of expansion and experimentation, the first Tower Hydroponic Units were developed in Armenia.

The Armenian tower hydroponicums are the first built examples of a vertical farm, and is documented in Sholto Douglas' seminal text "Hydroponics: The Bengal System" first published in 1951 with data from the then-East Pakistan, today's Bangladesh, and the Indian state of West Bengal. Contemporary notions of vertical farming are predated by this early technology by more than 50 years. Contemporary precursors that have been published, or built, are Ken Yeang's Bioclimatic Skyscraper (Menara Mesiniaga, built 1992); MVRDV's PigCity, 2000; MVRDV's Meta City/ Datatown (1998–2000); Pich-Aguilera's Garden Towers (2001).

Ken Yeang is perhaps the most widely known architects that has promoted the idea of the 'mixed-use' Bioclimatic Skyscraper which combines living units and opportunities for food production.

Early prototypes of vertical farms, or "Tower Hydroponicums" existed in Armenia prior to 1951 during an era of hydroponic and horticultural building system research fueled by space exploration and a transatlantic technology race.

The latest version of these very idea is Dickson Despommier's "The Vertical Farm".

Dickson Despommier, a professor of environmental health sciences and microbiology at Columbia University in New York City, modernized the idea of vertical farming in 1999 with graduate students in a medical ecology class. Although much of Despommier's suggestions have been challenged and strongly criticized from an environmental science and engineering point of view, the idea's popularization in recent years has been largely the result of Despommier's assertion that food production can be transformed.

Despommier had originally challenged his class to feed the entire population of Manhattan (About 2,000,000 people) using only 5 hectares (13 acres) of usable rooftop gardens. The class calculated that, by using rooftop gardening methods, only 2 percent of the population would be fed. Unsatisfied with the results, Despommier made an off-the-cuff suggestion of growing plants indoors, vertically. The idea sparked the students' interests and gained major momentum. By 2001 the first outline of a vertical farm was introduced and today scientists, architects, and investors worldwide are working together to make the concept of vertical farming a reality. In an interview with Miller-McCune.com, Despommier described how vertical farms would function:

> "Each floor will have its own watering and nutrient monitoring systems. There will be sensors for every single plant that tracks how much and what kinds of nutrients the plant has absorbed. You'll even have systems to monitor plant diseases by employing DNA chip technologies that detect the presence of plant pathogens by simply sampling the air and using snippets from various viral and bacterial infections. It's very easy to do.
>
> Moreover, a gas chromatograph will tell us when to pick the plant by analyzing which flavenoids the produce contains. These flavonoids are what gives the food the flavors you're so fond of, particularly for more aromatic produce like tomatoes and peppers. These are all right-off-the-shelf technologies. The ability to construct a vertical farm exists now. We don't have to make anything new."

Architectural designs have been produced by Chris Jacobs and Andrew Kranis from Columbia University and Gordon Graff from the University of Waterloo's School of Architecture in Cambridge, ON.

Mass media attention began with an article written in *New York* magazine. Since 2007, articles have appeared in *The New York Times*, *U.S. News & World Report*, *Popular Science*, *Scientific American* and *Maxim*, among others, as well as radio and television features.

As of 2012, Vertical Harvest is working on raising funds for an urban, small-scale vertical farm in Jackson Hole, Wyoming.

In 2013, July 18th, the Association for Vertical Farming was founded in Munich, Germany

As of 2014, Vertical Fresh Farms has been operating in Buffalo, NY specializing in a wide variety of salad greens, herbs, and sprouts.

Problems

Economics

Opponents question the potential profitability of vertical farming. A detailed cost analysis of start-up costs, operation costs and revenue has not been done. The extra cost of lighting, heating, and powering the vertical farm may negate any of the cost benefits received by the decrease in transportation expenses. The economic and environmental benefits of vertical farming rest partly on the concept of minimizing food miles, the distance that food travels from farm to consumer. However, a recent analysis suggests that transportation is only a minor contributor to the economic and environmental costs of supplying food to urban populations. The author of the report, University of Toronto professor Pierre Desrochers, concluded that "food miles are, at best, a marketing fad." Thus the facility would have to produce a considerable profit to justify remaining in the city. A simpler concept rather than trying to stack farms on multiple stories would be to just cultivate crops on the roofs of existing building. Rooftop farming is a growing urban trend, requires little construction (other than fortifying the roof to hold the weight of the growing medium), still takes advantage of sunlight and doesn't require investment in machinery, growing lights or irrigation.

Similarly, if the power needs of the vertical farm are met by fossil fuels, the environmental effect may be a net loss; even building low-carbon capacity to power the farms may not make as much sense as simply leaving the traditional farms in place, and burning less coal.

The initial building costs will be easily over $100 million, for a 60 hectare vertical farm. Office occupancy costs can be very high in major cities, with cities such as Tokyo, Moscow, Mumbai, Dubai, Milan, Zurich, and Sao Paulo ranging from $1850 to $880 per square meter, respectively.

Energy Use

During the growing season, the sun shines on a vertical surface at an extreme angle such that much less light is available to crops than when they are planted on flat land. Therefore, supplemental light would be required in order to obtain economically viable yields. Bruce Bugbee, a crop physiologist at Utah State University, believes that the power demands of vertical farming will be too expensive and uncompetitive with traditional farms using only free natural light. The environmental writer George Monbiot calculated that the cost of providing enough supplementary light to grow the grain for a single loaf would be about $15. An article in the Economist argued that "even though crops growing in a glass skyscraper will get some natural sunlight during the day, it won't be enough" and "the cost of powering artificial lights will make indoor farming prohibitively expensive".

As "The Vertical Farm" proposes a controlled environment, heating and cooling costs will be at least as costly as any other tower. But there also remains the issue of complicated, if not more expensive, plumbing and elevator systems to distribute food and water throughout. Even throughout the northern continental United States, while heating with relatively cheap fossil fuels, the heating cost can be over $200,000 per hectare.

To address this problem, The Plant in Chicago is building an anaerobic digester into the building. This will allow the farm to operate off the energy grid. Moreover, the anaerobic digester will be recycling waste from nearby businesses that would otherwise go into landfills.

Pollution

Depending on the method of electricity generation used, regular greenhouse produce can create more greenhouse gases than field produce, largely due to higher energy use per kilogram of produce. With vertical farms requiring much greater energy per kilogram of produce, mainly through increased lighting, than regular greenhouses, the amount of pollution created will be much higher than that from field produce. The amount of pollution produced is dependent on how the energy used in the process is generated.

As plants acquire nearly all their carbon from the atmosphere, greenhouse growers commonly supplement CO_2 levels to 3–4 times the rate normally found in the atmosphere. This increase in CO_2, which has been shown to increase photosynthesis rates by 50%, contributes to the higher yields expected in vertical farming. It is not uncommon to find greenhouses burning fossil fuels purely for this purpose, as other CO_2 sources, like from furnaces, contain pollutants such as sulphur dioxide and ethylene which significantly damage plants. This means a vertical farm will require a CO_2 source, most likely from combustion, even if the rest of the farm is powered by "green" energy. Also, through necessary ventilation, much CO_2 will be leaked into the city's atmosphere.

Greenhouse growers commonly exploit photoperiodism in plants to control whether the plants are in a vegetative or reproductive stage. As part of this control, growers will have the lights on past sunset and before sunrise or periodically throughout the night. Single story greenhouses are already a nuisance to neighbours because of light pollution, a 30-story vertical farm in a densely populated area will surely face problems because of its light pollution.

Hydroponics greenhouses regularly change the water, meaning there is a large quantity of water containing fertilizers and pesticides that must be disposed of. While solutions are currently being worked on, the most common method of simply spreading the mixture over a sufficient area of neighbouring farmland or wetlands would be more difficult for an urban vertical farm.

Advantages

Several potential advantages of vertical farming have been discussed by Despommier. Many of these benefits are obtained from scaling up hydroponic or aeroponic growing methods.

Preparation for the Future

It is estimated that by the year 2050, close to 80% of the world's population will live in urban areas and the total population of the world will increase by 3 billion people. A very large amount of land may be required depending on the change in yield per hectare. Scientists are concerned that this large amount of required farmland will not be available and that severe damage to the earth will be caused by the added farmland. According to Despommier, vertical farms, if designed properly, may eliminate the need to create additional farmland and help create a cleaner environment.

Increased Crop Production

Unlike traditional farming in non-tropical areas, indoor farming can produce crops year-round. All-season farming multiplies the productivity of the farmed surface by a factor of 4 to 6 depending on the crop. With some crops, such as strawberries, the factor may be as high as 30.

Furthermore, as the crops would be sold in the same infrastructures in which they are grown, they will not need to be transported between production and sale, resulting in less spoilage, infestation, and energy required than conventional farming encounters. Research has shown that 30% of harvested crops are wasted due to spoilage and infestation, though this number is much lower in developed nations.

Despommier suggests that, if dwarf versions of certain crops are used (e.g. dwarf wheat, which has been grown in space by NASA, is smaller in size but richer in nutrients), year-round crops, and "stacker" plant holders are accounted for, a 30-story building with a base of a building block (2 hectares (5 acres)) would yield a yearly crop analogous to that of 1,000 hectares (2,400 acres) of traditional farming.

Protection from Weather-related Problems

Crops grown in traditional outdoor farming suffer from the often suboptimal, and sometimes extreme, nature of geological and meteorological events such as undesirable temperatures or rainfall amounts, monsoons, hailstorms, tornadoes, flooding, wildfires, and severe droughts. The protection of crops from weather is increasingly important as global climate change occurs. "Three recent floods (in 1993, 2007 and 2008) cost the United States billions of dollars in lost crops, with even more devastating losses in topsoil. Changes in rain patterns and temperature could diminish India's agricultural output by 30 percent by the end of the century."

Because vertical plant farming provides a controlled environment, the productivity of vertical farms would be mostly independent of weather and protected from extreme weather events. Although the controlled environment of vertical farming negates most of these factors, earthquakes and tornadoes still pose threats to the proposed infrastructure, although this again depends on the location of the vertical farms.

Conservation of Resources

Each unit of area in a vertical farm could allow up to 20 units of area of outdoor farmland to return to its natural state, and recover farmlands due to development from original flat farmlands.

Vertical farming would reduce the need for new farmland due to overpopulation, thus saving many natural resources, currently threatened by deforestation or pollution. Deforestation and desertification caused by agricultural encroachment on natural biomes would be avoided. Because vertical farming lets crops be grown closer to consumers, it would substantially reduce the amount of fossil fuels currently used to transport and refrigerate farm produce. Producing food indoors reduces or eliminates conventional plowing, planting, and harvesting by farm machinery, also powered by fossil fuels.

Halting Mass Extinction

Withdrawing human activity from large areas of the Earth's land surface may be necessary to slow and eventually halt the current anthropogenic mass extinction of land animals.

Traditional agriculture is highly disruptive to wild animal populations that live in and around farmland and some argue it becomes unethical when there is a viable alternative. One study showed that wood mouse populations dropped from 25 per hectare to 5 per hectare after harvest, estimating 10 animals killed per hectare each year with conventional farming. In comparison, vertical farming would cause very little harm to wildlife, and would allow disused farmland to return to its pre-agricultural state.

Impact on Human Health

Traditional farming is a hazardous occupation with particular risks that often take their toll on the health of human laborers. Such risks include: exposure to infectious diseases such as malaria and schistosomes, exposure to toxic chemicals commonly used as pesticides and fungicides, confrontations with dangerous wildlife such as venomous snakes, and the severe injuries that can occur when using large industrial farming equipment. Whereas the traditional farming environment inevitably contains these risks (particularly in the farming practice known as "slash and burn"), vertical farming – because the environment is strictly controlled and predictable – reduces some of these dangers. Currently, the American food system makes fast, unhealthy food cheap while fresh produce is less available and more expensive, encouraging poor eating habits. These poor eating habits lead to health problems such as obesity, heart disease, and diabetes. The increased availability and subsequent lower cost of fresh produce would encourage healthier eating.

Poverty/Destitution and Culture

Food security is one of the primary factors leading to absolute poverty. Being able to construct 'farm land' in secure areas as needed will help alleviate the pressures causing crises among neighbors fighting for resources (mainly water and space). It also allows continued growth of culturally significant food items without sacrificing sustainability or basic needs, which can be significant to the recovery of a society from poverty.

Urban Growth

Vertical farming, used in conjunction with other technologies and socioeconomic practices, could allow cities to expand while remaining largely self-sufficient food wise. This would allow for large urban centers that could grow without destroying considerably larger areas of forest to provide food for their people. Moreover, the industry of vertical farming will provide employment to these expanding urban centers. This may help displace the unemployment created by the dismantling of traditional farms, as more farm laborers move to cities in search of work.

Energy Sustainability

Vertical farms could exploit methane digesters to generate a small portion of its own electrical needs. Methane digesters could be built on site to transform the organic waste generated at the farm into biogas which is generally composed of 65% methane along with other gases. This biogas could then be burned to generate electricity for the greenhouse.

Technologies and Devices

Vertical farming relies on the use of various physical methods to become effective. Combining these technologies and devices in an integrated whole is necessary to make Vertical Farming a reality. Various methods are proposed and under research. The most common technologies suggested are:

- Greenhouse
- The Folkewall and other vertical growing architectures
- Flowerpot
- Aeroponics / Hydroponics / Aquaponics
- Composting
- Grow light
- Phytoremediation
- Skyscraper
- Controlled-environment agriculture
- Precision agriculture
- Agricultural robot

Plans

Despommier argues that the technology to construct vertical farms currently exists. He also states that the system can be profitable and effective, a claim evidenced by some preliminary research posted on the project's website. Developers and local governments in the following cities have expressed serious interest in establishing a vertical farm: Incheon (South Korea), Abu Dhabi (United Arab Emirates), and Dongtan (China), New York City, Portland, Ore., Los Angeles, Las Vegas, Seattle, Surrey, B.C., Toronto, Paris, Bangalore, Dubai, Shanghai and Beijing. The Illinois Institute of Technology is now crafting a detailed plan for Chicago. It is suggested that prototype versions of vertical farms should be created first, possibly at large universities interested in the research of vertical farms, in order to prevent failures such as the Biosphere 2 project in Oracle, Arizona.

In 2009, the world's first pilot production system was installed at Paignton Zoo Environmental Park in the United Kingdom. The project showcased a technological solution for vertical farming and provided a physical base to conduct research into sustainable urban food production. The produce is used to feed the zoo's animals while the project enables evaluation of the systems and provides an educational resource to advocate for change in unsustainable land use practices that impact upon global biodiversity and ecosystem services,

In 2010, the Green Zionist Alliance proposed a resolution at the 36th World Zionist Congress calling on Keren Kayemet L'Yisrael (Jewish National Fund in Israel) to develop vertical farms in Israel.

In 2012, the world's first commercial vertical farm was opened in Singapore, developed by Sky Greens Farms, and is three stories high. They currently have over 100 towers that stand at nine meters tall.

In 2013 (July 18) the Association for Vertical Farming (AVF) was founded in Munich (Germany). By May 2015 the AVF already expanded with regional chapters all over Europe, Asia, USA, Canada and the United Kingdom. This internationally active non-profit organization unites growers and inventors to improve food security and the sustainable development of Vertical Farming. The AVF also focuses on advancing vertical farming technologies, designs and businesses by hosting international info-days, workshops and summits. They developed a glossary to bring consistency to the industry and plan on helping to standardize the technologies.

Greenhouse

A greenhouse (also called a glasshouse, or, if with sufficient heating, a hothouse) is a structure with walls and roof made chiefly of transparent material, such as glass, in which plants requiring regulated climatic conditions are grown. These structures range in size from small sheds to industrial-sized buildings. A miniature greenhouse is known as a cold frame. The interior of a greenhouse exposed to sunlight becomes significantly warmer than the external ambient temperature, protecting its contents in cold weather.

Victoria amazonica (giant Amazon waterlilies) in a large greenhouse at the Saint Petersburg Botanical Garden, Russia.

The Eden Project, in Cornwall, England.

The Royal Greenhouses of Laeken, Brussels, Belgium. An example of 19th-century greenhouse architecture

Many commercial glass greenhouses or hothouses are high tech production facilities for vegetables or flowers. The glass greenhouses are filled with equipment including screening installations, heating, cooling, lighting, and may be controlled by a computer to optimize conditions for plant growth.

Operation

The explanation given in most sources for the warmer temperature in a greenhouse is that incident solar radiation (the visible and adjacent portions of the infrared and ultraviolet ranges of the spectrum) passes through the glass roof and walls and is absorbed by the floor, earth, and contents, which become warmer and re-emit the energy as longer-wavelength infrared radiation. Glass and other materials used for greenhouse walls do not transmit infrared radiation, so the infrared cannot escape via radiative transfer. As the structure is not open to the atmosphere, heat also cannot escape via convection, so the temperature inside the greenhouse rises. This is known as the "greenhouse effect". The greenhouse effect, due to infrared-opaque "greenhouse gases", including carbon dioxide and methane instead of glass, also affects the earth as a whole; there is no convective cooling as air does not escape from the earth.

However, R. W. Wood in 1909 constructed two greenhouses, one with glass as the transparent material, and the other with panes of rock salt, which is transparent to infrared. The two greenhouses warmed to similar temperatures, suggesting that an actual greenhouse is warmer not because of the "greenhouse effect" as described in the previous paragraph, but by preventing convective cooling, not allowing warmed air to escape.

More recent quantitative studies suggest that the effect of infrared radiative cooling is not negligibly small, and may have economic implications in a heated greenhouse. Analysis of issues of near-infrared radiation in a greenhouse with screens of a high coefficient of reflection concluded that installation of such screens reduced heat demand by about 8%, and application of dyes to transparent surfaces was suggested. Composite less-reflective glass, or less effective but cheaper anti-reflective coated simple glass, also produced savings.

Types

Greenhouses can be divided into glass greenhouses and plastic greenhouses.

In domestic greenhouses, the glass used is typically 3mm (or 1/8″) 'horticultural glass' grade, which is good quality glass that should not contain air bubbles (which can produce scorching on leaves by acting like lenses).

Plastics mostly used are polyethylene film and multiwall sheets of polycarbonate material, or PMMA acrylic glass.

Commercial glass greenhouses are often high-tech production facilities for vegetables or flowers. The glass greenhouses are filled with equipment such as screening installations, heating, cooling and lighting, and may be automatically controlled by a computer.

Dutch Light

In the UK and other Northern European countries a pane of horticultural glass referred to as "Dutch Light" was historically used as a standard unit of construction, having dimensions of 28¾″ x 56″ (approx. 730mm x 1422 mm). This size gives a larger glazed area when compared with using smaller panes such as the 600mm width typically used in modern domestic designs which then require more supporting framework for a given overall greenhouse size. A style of greenhouse having sloped sides (resulting in a wider base than at eaves height) and using these panes uncut is also often referred to as of "Dutch Light design", and a cold frame using a full- or half-pane as being of "Dutch" or "half-Dutch" size.

Uses

Greenhouses allow for greater control over the growing environment of plants. Depending upon the technical specification of a greenhouse, key factors which may be controlled include temperature, levels of light and shade, irrigation, fertilizer application, and atmospheric humidity. Greenhouses may be used to overcome shortcomings in the growing qualities of a piece of land, such as a short growing season or poor light levels, and they can thereby improve food production in marginal environments. Greenhouses in hot, dry climates used specifically to provide shade are sometimes called "shadehouses".

As they may enable certain crops to be grown throughout the year, greenhouses are increasingly important in the food supply of high-latitude countries. One of the largest complexes in the world is in Almería, Andalucía, Spain, where greenhouses cover almost 200 km^2 (49,000 acres).

Greenhouses are often used for growing flowers, vegetables, fruits, and transplants. Special greenhouse varieties of certain crops, such as tomatoes, are generally used for commercial production. Many vegetables and flowers can be grown in greenhouses in late winter and early spring, and then transplanted outside as the weather warms. Bumblebees are the pollinators of choice for most pollination, although other types of bees have been used, as well as artificial pollination. Hydroponics can be used to make the most use of the interior space.

Greenhouses in Almería as seen from space. Credit: NASA

The relatively closed environment of a greenhouse has its own unique management requirements, compared with outdoor production. Pests and diseases, and extremes of heat and humidity, have to be controlled, and irrigation is necessary to provide water. Most greenhouses use sprinklers or drip lines. Significant inputs of heat and light may be required, particularly with winter production of warm-weather vegetables.

Greenhouses also have applications outside of the agriculture industry. GlassPoint Solar, located in Fremont, California, encloses solar fields in greenhouses to produce steam for solar-enhanced oil recovery.

Alpine House

An "alpine house" is a specialized greenhouse used for growing alpine plants. The purpose of an alpine house is to mimic the conditions in which alpine plants grow; particularly to provide protection from wet conditions in winter. Alpine houses are often unheated, since the plants grown there are hardy, or require at most protection from hard frost in the winter. They are designed to have excellent ventilation.

History

The idea of growing plants in environmentally controlled areas has existed since Roman times. The Roman emperor Tiberius ate a cucumber-like vegetable daily. The Roman gardeners used artificial methods (similar to the greenhouse system) of growing to have it available for his table every day of the year. Cucumbers were planted in wheeled carts which were put in the sun daily, then taken inside to keep them warm at night. The cucumbers were stored under frames or in cucumber houses glazed with either oiled cloth known as *specularia* or with sheets of selenite (a.k.a. *lapis specularis*), according to the description by Pliny the Elder.

Cucumbers reached to the ceiling in a greenhouse in Richfield, Minnesota, where market gardeners grew a wide variety of produce for sale in Minneapolis, *circa* 1910

19th-century orangerie in Weilburg, Germany

In the 13th century, greenhouses were built in Italy to house the exotic plants that explorers brought back from the tropics. They were originally called *giardini botanici* (botanical gardens).

'Active' greenhouses, in which it is possible for the temperature to be increased or decreased manually, appeared much later. *Sanga yorok*, written in the year 1450 AD in Korea, contained descriptions of a greenhouse which was designed to regulate the temperature and humidity requirements of plants and crops. One of the earliest records of the Annals of the Joseon Dynasty in 1438 confirms growing mandarin trees in a Korean traditional greenhouse during the winter and installing a heating system of ondol.

The concept of greenhouses also appeared in the Netherlands and then England in the 17th century, along with the plants. Some of these early attempts required enormous amounts of work to close up at night or to winterize. There were serious problems with providing adequate and balanced heat in these early greenhouses. Today, the Netherlands has many of the largest greenhouses in the world, some of them so vast that they are able to produce millions of vegetables every year.

The French botanist Charles Lucien Bonaparte is often credited with building the first practical modern greenhouse in Leiden, Holland, during the 1800s to grow medicinal tropical plants. Originally only on the estates of the rich, the growth of the science of botany caused greenhouses to spread to the universities. The French called their first greenhouses *orangeries*, since they were used to protect orange trees from freezing. As pineapples became popular, *pineries*, or pineapple pits, were built.

Experimentation with the design of greenhouses continued during the 17th century in Europe, as technology produced better glass and construction techniques improved. The greenhouse at the

Palace of Versailles was an example of their size and elaborateness; it was more than 150 metres (490 ft) long, 13 metres (43 ft) wide, and 14 metres (46 ft) high.

The golden era of the greenhouse was in England during the Victorian era, where the largest glass-houses yet conceived were constructed, as the wealthy upper class and aspiring botanists competed to build the most elaborate buildings. A good example of this trend is the pioneering Kew Gardens. Joseph Paxton, who had experimented with glass and iron in the creation of large greenhouses as the head gardener at Chatsworth, in Derbyshire, working for the Duke of Devonshire, designed and built The Crystal Palace in London, (although the latter was constructed for both horticultural and non-horticultural exhibition).

Other large greenhouses built in the 19th century included the New York Crystal Palace, Munich's Glaspalast and the Royal Greenhouses of Laeken (1874–1895) for King Leopold II of Belgium.

In Japan, the first greenhouse was built in 1880 by Samuel Cocking, a British merchant who exported herbs.

In the 20th century, the geodesic dome was added to the many types of greenhouses. Notable examples are the Eden Project, in Cornwall, The Rodale Institute in Pennsylvania, the Climatron at the Missouri Botanical Garden in St. Louis, Missouri, and Toyota Motor Manufacturing Kentucky.

Greenhouse structures adapted in the 1960s when wider sheets of polyethylene film became widely available. Hoop houses were made by several companies and were also frequently made by the growers themselves. Constructed of aluminum extrusions, special galvanized steel tubing, or even just lengths of steel or PVC water pipe, construction costs were greatly reduced. This resulted in many more greenhouses being constructed on smaller farms and garden centers. Polyethylene film durability increased greatly when more effective UV-inhibitors were developed and added in the 1970s; these extended the usable life of the film from one or two years up to 3 and eventually 4 or more years.

A plastic air-insulated greenhouse in New Zealand

Gutter-connected greenhouses became more prevalent in the 1980s and 1990s. These greenhouses have two or more bays connected by a common wall, or row of support posts. Heating inputs were reduced as the ratio of floor area to roof area was increased substantially. Gutter-connected greenhouses are now commonly used both in production and in situations where plants are grown and sold to the public as well. Gutter-connected greenhouses are commonly covered with structured polycarbonate materials, or a double layer of polyethylene film with air blown between to provide increased heating efficiencies.

Giant greenhouses in the Netherlands

Netherlands

The Netherlands has some of the largest greenhouses in the world. Such is the scale of food production in the country that in 2000, greenhouses occupied 10,526 hectares, or 0.25% of the total land area.

Greenhouses in the Westland region, Netherlands

Greenhouses began to be built in the Westland region of the Netherlands in the mid-19th century. The addition of sand to bogs and clay soil created fertile soil for agriculture, and around 1850, grapes were grown in the first greenhouses, simple glass constructions with one of the sides consisting of a solid wall. By the early 20th century, greenhouses began to be constructed with all sides built using glass, and they began to be heated. This also allowed for the production of fruits and vegetables that did not ordinarily grow in the area. Today, the Westland and the area around Aalsmeer have the highest concentration of greenhouse agriculture in the world. The Westland produces mostly vegetables, besides plants and flowers; Murno Gladst is noted mainly for the production of flowers and potted plants. Since the 20th century, the area around Venlo and parts of Drenthe have also become important regions for greenhouse agriculture.

Since 2000, technical innovations include the "closed greenhouse", a completely closed system allowing the grower complete control over the growing process while using less energy. Floating greenhouses are used in watery areas of the country.

Young tomatoes in an industrial-sized greenhouse in the Netherlands

The Netherlands has around 4,000 greenhouse enterprises that operate over 9,000 hectares of greenhouses and employ some 150,000 workers, producing €7.2 billion worth of vegetables, fruit, plants, and flowers, some 80% of which is exported.

Greenhouse Ventilation

Ventilation is one of the most important components in a successful greenhouse. If there is no proper ventilation, greenhouses and their growing plants can become prone to problems. The main purposes of ventilation are to regulate the temperature and humidity to the optimal level, and to ensure movement of air and thus prevent build-up of plant pathogens (such as *Botrytis cinerea*) that prefer still air conditions. Ventilation also ensures a supply of fresh air for photosynthesis and plant respiration, and may enable important pollinators to access the greenhouse crop.

Ventilation can be achieved via use of vents - often controlled automatically via a computer - and recirculation fans.

Improvements to natural ventilation systems result in an increase in production without significant cost compared with other climate control systems. The ventilation capacity of the Almería-type greenhouse must be improved, since it is still currently insufficient despite the improvements in recent years. The average window surface is 14.4%, which is far from the minimum window surface value of 30% that is required for correct ventilation or even the value of 25%, which is recommended in the Regulation for Integrated Production.

Greenhouse Heating

Heating or electricity is one of the most considerable costs in the operation of greenhouses across the globe, especially in colder climates. The main problem with heating a greenhouse as opposed to a building that has solid opaque walls is the amount of heat lost through the greenhouse covering. Since the coverings need to allow light to filter into the structure, they conversely cannot insulate very well. With traditional plastic greenhouse coverings having an R-value of around 2, a great amount of money is therefore spent to continually replace the heat lost. Most greenhouses, when supplemental heat is needed use natural gas or electric furnaces.

Passive heating methods exist which seek heat using low energy input. Solar energy can be captured from periods of relative abundance (day time/summer), and released to boost the temperature during cooler periods (night time/winter). Waste heat from livestock can also be used to heat greenhouses, e.g., placing a chicken coop inside a greenhouse recovers the heat generated by the chickens, which would otherwise be wasted.

Electronic controllers are often used to monitor the temperature and adjusts the furnace operation to the conditions. This can be as simple as a basic thermostat, but can be more complicated in larger greenhouse operations.

Greenhouse Carbon Dioxide Enrichment

The possibility of using carbon dioxide enrichment in greenhouse cultivation to enhance plant growth has been known for nearly 100 years. After the development of equipment for the controlled serial enrichment of carbon dioxide, the technique was established on a broad scale in the Netherlands. Secondary metabolites, e.g., cardiac glycosides in Digitalis lanata, are produced in higher amounts by greenhouse cultivation at enhanced temperature and at enhanced carbon dioxide concentration. Commercial greenhouses are now frequently located near appropriate industrial facilities for mutual benefit. For example, Cornerways Nursery in the UK is strategically placed near a major sugar refinery, consuming both waste heat and CO_2 from the refinery which would otherwise be vented to atmosphere. The refinery reduces its carbon emissions, whilst the nursery enjoys boosted tomato yields and does not need to provide its own greenhouse heating.

Enrichment only becomes effective where, by Liebig's law, carbon dioxide has become the limiting factor. In a controlled greenhouse, irrigation may be trivial, and soils may be fertile by default. In less-controlled gardens and open fields, rising CO_2 levels only increase primary production to the point of soil depletion (assuming no droughts, flooding, or both), as demonstrated *prima facie* by CO_2 levels continuing to rise. In addition, laboratory experiments, free air carbon enrichment (FACE) test plots, and field measurements provide replicability.

Plant Nursery

Plants in a nursery

A nursery is a place where plants are propagated and grown to usable size. They include retail nurseries which sell to the general public, wholesale nurseries which sell only to businesses such as other nurseries and to commercial gardeners, and private nurseries which supply the needs of institutions or private estates. Some retail and wholesale nurseries sell by mail.

Nurseries may supply plants for gardens, for agriculture, for forestry and for conservation biology.

Some nurseries specialize in one phase of the process: propagation, growing out, or retail sale; or in one type of plant: e.g., groundcovers, shade plants, or rock garden plants. Some produce bulk stock, whether seedlings or grafted, of particular varieties for purposes such as fruit trees for orchards, or timber trees for forestry. Some produce stock seasonally, ready in springtime for export to colder regions where propagation could not have been started so early, or to regions where seasonal pests prevent profitable growing early in the season.

Methods

Nurseries can grow plants in open fields, on container fields and in tunnels or greenhouses. In open fields nurseries grow ornamental trees, shrubs and herbaceous perennials, especially the plants meant for the wholesale trade or for amenity plantings. On a containerfield nurseries grow small trees, shrubs and herbaceous plants, usually destined for sales in garden centers. Nurseries also grow plants in greenhouses, a building of glass or in plastic tunnels, designed to protect young plants from harsh weather (especially frost), while allowing access to light and ventilation. Modern greenhouses allow automated control of temperature, ventilation and light and semi-automated watering and feeding. Some also have fold-back roofs to allow "hardening-off" of plants without the need for manual transfer to outdoor beds.

An Orchid nursery

A tree nursery using gutters to decrease growing costs

Most nurseries remain highly labor-intensive. Although some processes have been mechanised and automated, others have not. It remains highly unlikely that all plants treated in the same way at the same time will arrive at the same condition together, so plant care requires observation, judgment and manual dexterity; selection for sale requires comparison and judgment. A UK nurseryman has estimated (in 2003) that manpower accounts for 70% of his production costs. The largest UK nurseries have moved to minimize labour costs by the use of computer controlled warehousing methods: plants are pallet allocated to a location and grown on there with little human intervention. Picking merely requires selection of a batch and manual quality control before dispatch. In other cases, a high loss rate during maturation is accepted for the reduction in detailed plant maintenance costs.

Business is highly seasonal, concentrated in spring and autumn. There is no guarantee that there will be demand for the product - this will be affected by temperature, drought, cheaper foreign competition, fashion, among other things.

Annuals are sold in trays (undivided containers with multiple plants), flats (trays with built-in cells), peat pots, or plastic pots. Perennials and woody plants are sold either in pots, bare-root or balled and burlapped and in a variety of sizes, from liners to mature trees. Balled and Burlap (B & B) trees are dug either by hand or by a loader that has a tree spade attachment on the front of the machine. Although container grown woody plants are becoming more and more popular due to the versatility, B & B is still widely used throughout the industry.

Plants may be propagated by seeds, but often desirable cultivars are propagated asexually. The most common method is by cuttings. These can be taken from shoot tips or parts of stems with a node (softwood cuttings) or from older stems (hardwood cuttings). Herbaceous perennials are also often propagated by root cuttings or division. For plants on a rootstock grafting or budding is used. Older techniques like layering are sometimes used for crops which are difficult to propagate.

Conditioning

With the objective of fitting planting stock more ably to withstand stresses after outplanting, various nursery treatments have been attempted or developed and applied to nursery stock. Buse and Day (1989), for instance, studied the effect of conditioning of white spruce and black spruce transplants on their morphology, physiology, and subsequent performance after outplanting. Root pruning, wrenching, and fertilization with potassium at 375 kg/ha were the treatments applied. Root pruning and wrenching modified stock in the nursery by decreasing height, root collar diameter, shoot:root ratio, and bud size, but did not improve survival or growth after planting. Fertilization reduced root growth in black spruce but not of white spruce.

Hardening off, Frost Hardiness

Seedlings vary in their susceptibility to injury from frost. Damage can be catastrophic if "unhardened" seedlings are exposed to frost. Frost hardiness may be defined as the minimum temperature at which a certain percentage of a random seedling population will survive or will sustain a given level of damage (Siminovitch 1963, Timmis and Worrall 1975). The term LT_{50} (lethal temperature for 50% of a population) is commonly used. Determination of frost hardiness in Ontario is based on electrolyte leakage from mainstem terminal tips 2 cm to 3 cm long in weekly samplings (Colombo and Hickie 1987). The tips are frozen then thawed, immersed in distilled water, the electrical conductivity of which depends on the degree to which cell membranes have been ruptured by freezing releasing electrolyte. A -15 °C frost hardiness level has been used to determine the readiness of container stock to be moved outside from the greenhouse, and -40 °C has been the level determining readiness for frozen storage (Colombo 1997).

In an earlier technique, potted seedlings were placed in a freezer chest and cooled to some level for some specific duration; a few days after removal, seedlings were assessed for damage using various criteria, including odour, general visual appearance, and examination of cambial tissue (Ritchie 1982).

Stock for fall planting must be properly hardened-off. Conifer seedlings are considered to be hardened off when the terminal buds have formed and the stem and root tissues have ceased growth. Other characteristics that in some species indicate dormancy are color and stiffness of the needles, but these are not apparent in white spruce.

Forest Tree Nurseries

Whether in the forest or in the nursery, seedling growth is fundamentally influenced by soil fertility, but nursery soil fertility is readily amenable to amelioration, much more so than is forest soil.

Nitrogen, phosphorus, and potassium are regularly supplied as fertilizers, and calcium and magnesium are supplied occasionally. Applications of fertilizer nitrogen do not build up in the soil to develop any appreciable storehouse of available nitrogen for future crops. Phosphorus and potassium, however, can be accumulated as a storehouse available for extended periods.

Fertilization permits seedling growth to continue longer through the growing season than unfertilized stock; fertilized white spruce attained twice the height of unfertilized. High fertility in the rooting medium favours shoot growth over root growth, and can produce top-heavy seedlings ill suited to the rigors of the outplant site. Nutrients in oversupply can reduce growth or the uptake of other nutrients. As well, an excess of nutrient ions can prolong or weaken growth to interfere with the necessary development of dormancy and hardening of tissues in time to withstand winter weather.

Stock types, Sizes and Lots

Nursery stock size typically follows the normal curve when lifted for planting stock. The runts at the lower end of the scale are usually culled to an arbitrary limit, but, especially among bareroot stock, the range in size is commonly considerable. Dobbs (1976) and McMinn (1985a) examined how the performance of 2+0 bareroot white spruce related to differences in initial size of planting stock. The stock was regraded into large, medium, and small fractions according to fresh weight. The small fraction (20% of the original stock) had barely one-quarter of the dry matter mass of the large fraction at the time of outplanting. Ten years later, in the blade-scarified site, seedlings of the large fraction had almost 50% greater stem volume than had seedlings of the small fraction. Without site preparation, large stock were more than twice the size of small stock after 10 years.

Similar results were obtained with regraded 2+1 transplants sampled to determine root growth capacity. The large stock had higher RGC as well as greater mass than the small stock fraction.

The value of large size at the time of planting is especially apparent when outplants face strong competition from other vegetation, although high initial mass does not guarantee success. That the growth potential of planting stock depends on much more than size seems clear from the indifferent success of the transplanting of small 2+0 seedlings for use as 2+1 "reclaim" transplants. The size of bareroot white spruce seedlings and transplants also had a major influence on field performance.

The field performance among various stock types in Ontario plantations was examined by Paterson and Hutchison (1989): the white spruce stock types were 2+0, 1.5+0.5, 1.5+1.5, and 3+0. The nursery stock was grown at Midhurst Forest Tree Nursery, and carefully handled through lifting on 3 lift dates, packing, and hot-planting into cultivated weed-free loam. After 7 years, overall survival was 97%, with no significant differences in survival among stock types. The 1.5+1.5 stock with a mean height of 234 cm was significantly taller by 18% to 25% than the other stock types. The 1.5+1.5 stock also had significantly greater dbh than the other stock types by 30-43%. The best stock type was 57 cm taller and 1 cm greater in dbh than the poorest. Lifting date had no significant effect on growth or survival.

High elevation sites in British Columbia's southern mountains are characterized by a short growing season, low air and soil temperatures, severe winters, and deep snow. The survival and growth of Engelmann spruce and subalpine fir outplanted in 3 silvicultural trials on such sites in gaps of various sizes were compared by Lajzerowicz et al. (2006). Survival after 5 or 6 years decreased with smaller gaps. Height and diameter also decreased with decreasing size of gap; mean heights were 50 cm to 78 cm after 6 years, in line with height expectations for Engelmann spruce in a high-elevation planting study in southeastern British Columbia. In the larger gaps (≥1.0 ha), height increment by year 6 was ranging from 10 cm to 20 cm. Lajzerrowicz et al. Concluded that plantings of conifers in clearcuts at high elevations in the southern mountains of British Columbia are likely to be successful, even close to timberline; and group selection silvicultural systems based on gaps 0.1 ha or larger are also likely to succeed. Gaps smaller than 0.1 ha do not provide suitable conditions for obtaining adequate survival or for growth of outplanted conifers.

Planting Stock

Planting stock, "seedlings, transplants, cuttings, and occasionally wildings, for use in planting out," is nursery stock that has been made ready for outplanting.The amount of seed used in white spruce seedling production and direct seeding varies with method.

A working definition of planting stock quality was accepted at the 1979 IUFRO *Workshop on Techniques for Evaluating Planting Stock Quality* in New Zealand: "The quality of planting stock is the degree to which that stock realizes the objectives of management (to the end of the rotation or achievement of specified sought benefits) at minimum cost. Quality is fitness for purpose." Clear expression of objectives is therefore prerequisite to any determination of planting stock quality. Not only does performance have to be determined, but performance has to be rated against the objectives of management. Planting stock is produced in order to give effect to the forest policy of the organization.

A distinction needs to be made between "planting stock quality" and "planting stock performance potential" (PSPP). The actual performance of any given batch of outplanted planting stock is determined only in part by the kind and condition, i.e., the intrinsic PSPP, of the planting stock.

The PSPP is impossible to estimate reliably by eye because outward appearance, especially of stock withdrawn from refrigerated storage, can deceive even experienced foresters, who would be offended if their ability were questioned to recognize good planting stock when they saw it. Prior to Wakeley's (1954) demonstration of the importance of the physiological state of planting stock in determining the ability of the stock to perform after outplanting, and to a considerable extent even afterwards, morphological appearance has generally served as the basis for estimating the quality of planting stock. Gradually, however, a realization developed that more was involved. Tucker et al. (1968), for instance, after assessing 10-year survival data from several experimental white spruce plantations in Manitoba noted that "Perhaps the most important point revealed here is that certain lots of transplants performed better than others", even though all transplants were handled and planted with care. The intuitive "stock that looks good must be good" is a persuasive, but potentially dangerous maxim. That greatest of teachers, Bitter Experience, has often enough demonstrated the fallibility of such assessment, even though the corollary "stock that looks bad must be bad" is likely to be well founded. The physiological qualities of planting stock are hidden from the eye and must be revealed by testing. The potential for survival and growth of a batch of planting stock may be estimated from various features, morphological and physiological, of the stock or a sample thereof.

The size and shape and general appearance of a seedling can nevertheless give useful indications of PSPP. In low-stress outplanting situations, and with a minimized handling and lifting-planting cycle, a system based on specification for nursery stock and minimum morphological standards for acceptable seedlings works tolerably well. In certain circumstances, benefits often accrue from the use of large planting stock of highly ranked morphological grades. Length of leading shoot, diameter of stem, volume of root system, shoot:root ratios, and height:diameter ratios have been correlated with performance under specific site and planting conditions. However, the concept that larger is better negates the underlying complexities. Schmidt-Vogt (1980), for instance, found that whereas mortality among large outplants is greater than among small in the year of planting, mortality in subsequent growing seasons is higher among small outplants than among large. Much of the literature on comparative seedling performance is clouded by uncertainty as to whether the stocks being compared share the same physiological condition; differences invalidate such comparisons.

Height and root-collar diameter are generally accepted as the most useful morphological criteria and are often the only ones used in specifying standards.Quantification of root system morphology is difficult but can be done, e.g. by using the photometric rhizometer to determine intercept area, or volume by displacement or gravimetric methods.

Planting stock is always subject to a variety of conditions that are never optimal *in toto*. The effect of sub-optimal conditions is to induce stress in the plants. The nursery manager aims, and is normally able to avoid stresses greater than moderate, i.e., restricting stresses to levels that can be tolerated by the plants without incurring serious damage. The adoption of nursery regimes to equip planting stock with characteristics conferring increased ability to withstand outplanting stresses, by managing stress levels in the nursery to "condition" planting stock to increase tolerance to various post-planting environmental stresses, has become widespread, particularly with containerized stock.

Outplanted stock that is unable to tolerate high temperatures occurring at soil surfaces will fail to establish on many forest sites, even in the far north. Factors affecting heat tolerance were investigated by Colombo et al. (1995); the production and roles of heat shock proteins (HSPs) are important in this regard. HSPs, present constitutively in black spruce and many other, perhaps most, higher plants are important both for normal cell functioning and in a stress response mechanism following exposure to high, non-lethal temperature. In black spruce at least, there is an association between HSPs and increased levels of heat tolerance. Investigation of the diurnal variability in heat tolerance of roots and shoots in black spruce seedlings 14 to 16 weeks old found in all 4 trials that shoot heat tolerance was significantly greater in the afternoon than in the morning. The trend in root heat tolerance was similar to that found in the shoots; root systems exposed to 47 °C for 15 minutes in the afternoon averaged 75 new roots after a 2-week growth period, whereas only 28 new roots developed in root systems similarly exposed in the morning. HSP_{73} was detected in black spruce nuclear, mitochondrial, microsomal, and soluble protein fractions, while HSP_{72} was observed only in the soluble protein fraction. Seedlings exhibited constitutive synthesis of HSP_{73} at 26 °C in all except the nuclear membrane fraction in the morning; HSP levels at 26 °C in the afternoon were higher than in the morning in the mitochondrial and microsomal protein factions. Heat shock affected the abundance of HSPs depending on protein fraction and time of day. Without heat shock, nuclear membrane-bound HSP_{73} was absent from plants in the morning and only weakly present in the afternoon, and heat shock increased the abundance of nuclear membrane. Heat shock also affected the abundance of HSP_{73} in the afternoon, and caused HSP_{73} to appear in the morning. In the mitochondrial and microsomal protein fractions, an afternoon heat shock reduced HSP_{73}, whereas a morning heat shock increased HSP_{73} in the mitochondrial but decreased it in the microsomal fraction. Heat shock increased soluble $HSP_{72/73}$ levels in both the morning and afternoon. In all instances, shoot and root heat tolerances were significantly greater in the afternoon than in the morning.

Planting stock continues to respire during storage even if frozen. Temperature is the major factor controlling the rate, and care must be taken to avoid overheating. Navratil (1982) found that closed containers in cold storage averaged internal temperatures 1.5 °C to 2.0 °C above the nominal storage temperature. Depletion of reserves can be estimated from the decrease in dry weight. Cold-stored 3+0 white spruce nursery stock in northern Ontario had lost 9% to 16% of dry weight after 40 days of storage. Carbohydrates can also be determined directly.

The propensity of a root system to develop new roots or extend existing roots cannot be determined by eye, yet it is the factor that makes or breaks the outcome of an outplanting operation. The post-planting development of roots or root systems of coniferous planting stock is determined by many factors, some physiological, some environmental. Unsatisfactory rates of post-planting survival unrelated to the morphology of the stock, led to attempts to test the physiological condition of planting stock, particularly to quantify the propensity to produce new root growth. New root growth can be assumed to be necessary for successful establishment of stock after planting, but although the thesis that RGC is positively related to field performance would seem to be reasonable, supporting evidence has been meager.

The physiological condition of seedlings is reflected by changes in root activity. This is helpful in determining the readiness of stock for lifting and storing and also for outplanting after storage. Navratil (1982) reported a virtually perfect ($R^2 = 0.99$) linear relationship in the frequency of 3+0 white spruce white root tips longer than 10 mm with time in the fall at Pine Ridge Forest Nursery,

Alberta, decreasing during a 3-week period to zero on October 13 in 1982.Root regenerating research with white spruce in Canada (Hambly 1973, Day and MacGillivray 1975, Day and Breunig 1997) followed similar lines to that of Stone's (1955) pioneering work in California.

Simpson and Ritchie (1997) debated the proposition that root growth potential of planting stock predicts field performance; their conclusion was that root growth potential, as a surrogate for seedling vigor, can predict field performance, but only under such situations as site conditions permit. Survival after planting is only partly a function of an outplant's ability to initiate roots in test conditions; root growth capacity is not the sole predictor of plantation performance.

Some major problems militate against greater use of RGC in forestry, including: unstandardized techniques; unstandardized quantification; uncertain correlation between quantified RGC and field performance; variability within given, nominally identical, kinds of planting stock; and the irrelevance of RGC test values determined on a sub-sample of a parent population that subsequently, before it is planted, undergoes any substantive physiological or physical change. In its present form, RGC testing is silviculturally useful chiefly as a means of detecting planting stock that, while visually unimpaired, is moribund.

Seedling moisture content can be increased or decreased in storage, depending on various factors including especially the type of container and the kind and amount of moisture-retaining material present. When seedlings exceed 20 bars PMS in storage, survival after outplanting becomes problematical. The Relative Moisture Content of stock lifted during dry conditions can be increased gradually when stored in appropriate conditions. White spruce (3+0) packed in Kraft bags in northern Ontario increased RMC by 20% to 36% within 40 days.

Bareroot 1.5+1.5 white spruce were taken from cold storage and planted early in May on a clear-felled boreal forest site in northeastern Ontario. Similar plants were potted and kept in a greenhouse. In outplanted trees, maximum stomatal conductances (g) were initially low (<0.01 cm/s), and initial base xylem pressure potentials (PSIb) were -2.0 MPa. During the growing season, g increased to about 0.20 cm/s and PSIb to -1.0 MPa. Minimum xylem pressure potential (PSIm) was initially -2.5 MPa, increasing to -2.0 MPa on day 40, and about -1.6 MPa by day 110. During the first half of the growing season, PSIm was below turgor loss point. The osmotic potential at turgor loss point decreased after planting to -2.3 MPa 28 days later. In the greenhouse, minimum values of PSIT were -2.5 MPa (in the first day after planting. the maximum bulk modulus of elasticity was greater in white spruce than in similarly treated jack pine and showed greater seasonal changes. Relative water content (RWC) at turgor loss was 80-87%. Available turgor (TA), defined as the integral of turgor over the range of RWC between PSIb and xylem pressure potential at the turgor loss point) was 4.0% for white spruce at the beginning of the season compared with 7.9% for jack pine, but for the rest of the season TA for jack pine was only 2%, to 3% that of white spruce. Diurnal turgor (Td), the integral of turgor over the range of RWC between PSIb and PSIm, as a percentage of TA was higher in field-planted white spruce than jack pine until the end of the season.

The stomata of both white and black spruce were more sensitive to atmospheric evaporative demands and plant moisture stress during the first growing season after outplanting on 2 boreal sites in northern Ontario than were jack pine stomata, physiological differences that favoured growth and establishment being more in jack pine than in the spruces.

With black spruce and jack pine, but not with white spruce, Grossnickle and Blake's (1987) findings warrant mention in relation to the bareroot-containerized debate. During the first growing season after outplanting, containerized seedlings of both species had greater needle conductance than bareroot seedlings over a range of absolute humidity deficits. Needle conductance of containerized seedlings of both species remained high during periods of high absolute humidity deficits and increasing plant moisture stress. Bareroot outplants of both species had a greater early season resistance to water-flow through the soil–plant–atmosphere continuum (SPAC) than had containerized outplants. Resistance to water flow through the SPAC decreased in bareroot stock of both species as the season progressed, and was comparable to containerized seedlings 9 to 14 weeks after planting. Bareroot black spruce had greater new-root development than containerized stock throughout the growing season.

The greater efficiency of water use in newly transplanted 3-year-old white spruce seedlings under low levels of absolute humidity difference in water-stressed plants immediately after planting helps explain the commonly observed favourable response of young outplants to the nursing effect of a partial canopy. Silvicultural treatments promoting higher humidity levels at the planting microsite should improve white spruce seedling photosynthesis immediately after planting.

Stock Types (Seedling Nomenclature)

Planting stock is grown under many diverse nursery culture regimes, in facilities ranging from sophisticated computerized greenhouses to open compounds. Types of stock include bareroot seedlings and transplants, and various kinds of containerized stock. For simplicity, both container-grown and bareroot stock are generally referred to as seedlings, and transplants are nursery stock that have been lifted and transplanted into another nursery bed, usually at wider spacing. The size and physiological character of stock vary with the length of growing period and with growing conditions. Until the technology of raising containerized nursery stock bourgeoned in the second half of the twentieth-century, bareroot planting stock classified by its age in years was the norm.

Classification by Age

The number of years spent in the nursery seedbed by any particular lot of planting stock is indicated by the 1st of a series of numbers. The 2nd number indicates the years subsequently spent in the transplant line, and a zero is shown if indeed there has been no transplanting. A 3rd number, if any, would indicate the years subsequently spent after a second lifting and transplanting. The numbers are sometimes separated by dashes, but separation by plus sign is more logical inasmuch as the sum of the individual numbers gives the age of the planting stock. Thus 2+0 is 2-year-old seedling planting stock that has not been transplanted, and Candy's (1929) white spruce 2+2+3 stock had spent 2 years in the seedbed, 2 years in transplant lines, and another 3 years in transplant lines after a second transplanting. Variations have included such self-explanatory combinations, such as 1½+1½, etc.

The class of planting stock to use on a particular site is generally selected on the basis of historical record of survival, growth, and total cost of surviving trees. In the Lake States, Kittredge concluded that good stock of 2+1 white spruce was the smallest size likely to succeed and was better than larger and more expensive stock when judged by final cost of surviving trees.

Classification by Seedling Description Code

Because age alone is an inadequate descriptor of planting stock, various codes have been developed to describe such components of stock characteristics as height, stem diameter, and shoot:-root ratio. A description code may include an indication of the intended planting season.

Physiological Characteristics

Neither age classification nor seedling description code indicate the physiological condition of planting stock, though rigid adherence to a given cultural regime together with observation of performance over a number of years of planting can produce stock suitable for performing on a "same again" basis.

Classification by System

Planting stock is raised under a variety of systems, but these have devolved generally into 2 main groupings: bareroot and containerized. Manuals specifically for the production of bareroot and containerized nursery stock are valuable resources for the nursery manager. As well, a lot of good information about nursery stock specific to regional jurisdictions is well presented by Cleary et al. (1978) for Oregon, Lavender et al. (1990) for British Columbia, and Wagner and Colombo (2001) for Ontario.

Plant Propagation

Plant propagation is the process of creating new plants from a variety of sources: seeds, cuttings, bulbs and other plant parts. Plant propagation can also refer to the artificial or natural dispersal of plants.

Gentian seedlings in a plant nursery

Sexual Propagation (Seed)

Seeds and spores can be used for reproduction (through e.g. sowing). Seeds are typically produced from sexual reproduction within a species, because genetic recombination has occurred. A plant grown from seeds may have different characteristics from its parents. Some species produce seeds that require special conditions to germinate, such as cold treatment. The seeds of many Australian plants and plants from southern Africa and the American west require smoke or fire to germinate. Some plant species, including many trees do not produce seeds until they reach maturity, which may take many years. Seeds can be difficult to acquire and some plants do not produce seed at all. Some plants (like certain F1/F2 hybrids and GMO plants) may produce seed, but not fertile seed. In certain cases (like with GMO's), this is done to prevent the accidental spreading of these plants (which are generally non-native crops), for example by birds and other animals.

One way to propagate an avocado seed

Asexual Propagation

Softwood stemcuttings rooting in a controlled environment

Plants have a number of mechanisms for asexual or vegetative reproduction. Some of these have been taken advantage of by horticulturists and gardeners to multiply or clone plants rapidly. People also use methods that plants do not use, such as tissue culture and grafting. Plants are produced using material from a single parent and as such there is no exchange of genetic material, therefore vegetative propagation methods almost always produce plants that are identical to the parent. Vegetative reproduction uses plants parts such as roots, stems and leaves. In some plants seeds can be produced without fertilization and the seeds contain only the genetic material of the parent plant. Therefore, propagation via asexual seeds or apomixis is asexual reproduction but not vegetative propagation.

Techniques for vegetative propagation include:

- Air or ground layering
- Division
- Grafting and bud grafting, widely used in fruit tree propagation
- Micropropagation
- Stolons or runners
- Storage organs such as bulbs, corms, tubers and rhizomes
- Striking or cuttings
- Twin-scaling
- Offsets

Heated Propagator

A heated propagator is a horticultural device to maintain a warm and damp environment for seeds and cuttings to grow in.

This can be in the form of a clear enclosed bin sitting over a hotpad, or even a portable heater pointed at the bin. The key is to keep the moisture in the clear bin, while keeping lighting over the top of it, usually.

Seed Propagation Mat

An electric seed-propagation mat is a heated rubber mat covered by a metal cage which is used in gardening. The mats are made so that planters containing seedlings can be placed on top of the metal cage without the risk of starting a fire. In extreme cold, gardeners place a loose plastic cover over the planters/mats which creates a sort of miniature greenhouse. The constant and predictable heat allows people to garden in the winter months when the weather is generally too cold for seedlings to survive naturally. When combined with a lighting system, many plants can be grown indoors using these mats.

Grafting

Grafting or graftage is a horticultural technique whereby tissues of plants are joined so as to continue their growth together. The upper part of the combined plant is called the scion while the lower part is called the rootstock. The success of this joining requires that the vascular tissue grow together and such joining is called inosculation. The technique is most commonly used in asexual propagation of commercially grown plants for the horticultural and agricultural trades.

Cherry tree, consolidated "V" graft

Tape has been used to bind the rootstock and scion at the graft and tar of the scion from desiccation.

A grafted tree showing two different color blossoms

In most cases, one plant is selected for its roots and this is called the stock or rootstock. The other plant is selected for its stems, leaves, flowers, or fruits and is called the scion or cion. The scion contains the desired genes to be duplicated in future production by the stock/scion plant.

In stem grafting, a common grafting method, a shoot of a selected, desired plant cultivar is grafted onto the stock of another type. In another common form called bud grafting, a dormant side bud is grafted onto the stem of another stock plant, and when it has inosculated successfully, it is encouraged to grow by pruning off the stem of the stock plant just above the newly grafted bud.

For successful grafting to take place, the vascular cambium tissues of the stock and scion plants must be placed in contact with each other. Both tissues must be kept alive until the graft has "taken", usually a period of a few weeks. Successful grafting only requires that a vascular connection take place between the grafted tissues. Joints formed by grafting are not as strong as naturally formed joints, so a physical weak point often still occurs at the graft because only the newly formed tissues inosculate with each other. The existing structural tissue (or wood) of the stock plant does not fuse.

Advantages

Graft particular to plum Cherry. The scion is the largest in the plant, due to the imperfect union of the two. It can be seen on the enlarged trunk: this accumulation of starch is an indication of imperfection.

- Precocity: The ability to induce fruitfulness without the need for completing the juvenile phase. Juvenility is the natural state through which a seedling plant must pass before it can become reproductive. In most fruiting trees, juvenility may last between 5 and 9 years, but in some tropical fruits e.g. Mangosteen, juvenility may be prolonged for up to 15 years. Grafting of mature scions onto rootstocks can result in fruiting in as little as two years.
- Dwarfing: To induce dwarfing or cold tolerance or other characteristics to the scion. Most apple trees in modern orchards are grafted on to dwarf or semi-dwarf trees planted at high density. They provide more fruit per unit of land, higher quality fruit, and reduce the danger of accidents by harvest crews working on ladders. Care must be taken when planting dwarf or semi-dwarf trees. If such a tree is planted with the graft below the soil, then the scion portion can also grow roots and the tree will still grow to its standard size.

- Ease of propagation: Because the scion is difficult to propagate vegetatively by other means, such as by cuttings. In this case, cuttings of an easily rooted plant are used to provide a rootstock. In some cases, the scion may be easily propagated, but grafting may still be used because it is commercially the most cost-effective way of raising a particular type of plant.
- Hybrid breeding: To speed maturity of hybrids in fruit tree breeding programs. Hybrid seedlings may take ten or more years to flower and fruit on their own roots. Grafting can reduce the time to flowering and shorten the breeding program.
- Hardiness: Because the scion has weak roots or the roots of the stock plants are tolerant of difficult conditions. e.g. many showy Western Australian plants are sensitive to dieback on heavy soils, common in urban gardens, and are grafted onto hardier eastern Australian relatives. Grevilleas and eucalypts are examples.
- Sturdiness: To provide a strong, tall trunk for certain ornamental shrubs and trees. In these cases, a graft is made at a desired height on a stock plant with a strong stem. This is used to raise 'standard' roses, which are rose bushes on a high stem, and it is also used for some ornamental trees, such as certain weeping cherries.
- Disease/Pest Resistance: In areas where soil-borne pests or pathogens would prevent the successful planting of the desired cultivar, the use of pest/disease tolerant rootstocks allow the production from the cultivar that would be otherwise unsuccessful. A major example is the use of rootstocks in combating Phylloxera.
- Pollen source: To provide pollenizers. For example, in tightly planted or badly planned apple orchards of a single variety, limbs of crab apple may be grafted at regularly spaced intervals onto trees down rows, say every fourth tree. This takes care of pollen needs at blossom time, yet does not confuse pickers who might otherwise mix varieties while harvesting, as the mature crab apples are so distinct from other apple varieties.
- Repair: To repair damage to the trunk of a tree that would prohibit nutrient flow, such as stripping of the bark by rodents that completely girdles the trunk. In this case a bridge graft may be used to connect tissues receiving flow from the roots to tissues above the damage that have been severed from the flow. Where a water-shoot, basal shoot or sapling of the same species is growing nearby, any of these can be grafted to the area above the damage by a method called inarch grafting. These alternatives to scions must be of the correct length to span the gap of the wound.
- Changing cultivars: To change the cultivar in a fruit orchard to a more profitable cultivar, called *top working*. It may be faster to graft a new cultivar onto existing limbs of established trees than to replant an entire orchard.
- Maintain consistency: Apples are notorious for their genetic variability, even differing in multiple characteristics, such as, size, color, and flavor, of fruits located on the same tree. In the commercial farming industry, consistency is maintained by grafting a scion with desired fruit traits onto a hardy stock.

An example of approach grafting by Axel Erlandson.

- Curiosities
  - A practice sometimes carried out by gardeners is to graft related potatoes and tomatoes so that both are produced on the same plant, one above ground and one underground.
  - Cacti of widely different forms are sometimes grafted on to each other.
  - Multiple cultivars of fruits such as apples are sometimes grafted on a single tree. This so-called "family tree" provides more fruit variety for small spaces such as a suburban backyard, and also takes care of the need for pollenizers. The drawback is that the gardener must be sufficiently trained to prune them correctly, or one strong variety will usually "take over."
  - Ornamental and functional, tree shaping uses grafting techniques to join separate trees or parts of the same tree to itself. Furniture, hearts, entry archways are examples. Axel Erlandson was a prolific tree shaper who grew over 75 mature specimens.

Techniques

Approach

T budding

Approach grafting or inarching is used to join together plants that are otherwise difficult to join. The plants are grown close together, and then joined so that each plant has roots below and growth above the point of union. Both scion and stock retain their respective parents that may or may not be removed after joining. Also used in pleaching. The graft can be successfully accomplished any time of year.

Budding

Bud grafting uses a bud instead of a twig. Grafting roses is the most common example of bud grafting. In this method a bud is removed from the parent plant, and the base of the bud is inserted beneath the bark of the stem of the stock plant from which the rest of the shoot has been cut. Any extra bud that starts growing out from the stem of the stock plant is removed because that would bear the flower of the unwanted original kind. Examples: roses and peaches. Budding is the easiest technique used.

Cleft

In cleft grafting a small cut is made in the stalk and then the pointed end of the scion is inserted in the stalk. The most common form of grafting is cleft grafting. This is best done in the spring and is useful for joining a thin scion about 1 cm (0.39 in) diameter to a thicker branch or stock. It is best if the latter is 2–7 cm (0.79–2.76 in) in diameter and has 3–5 buds. The branch or stock should be split carefully down the middle to form a cleft about 3 cm (1.2 in) deep. If it is a branch that is not vertical then the cleft should be cut horizontally. The end of the scion should be cut cleanly to a long shallow wedge, preferably with a single cut for each wedge surface, and not whittled. A third cut may be made across the end of the wedge to make it straight across.

Successful cleft graft after 2 years' growth

Same graft after 4 years' growth. Graft is now final, rootstock completely overtaken

Slide the wedge into the cleft so that it is at the edge of the stock and the centre of the wedge faces are against the cambium layer between the bark and the wood. It is preferable if a second scion is inserted in a similar way into the other side of the cleft. This helps to seal off the cleft. Tape around the top of the stock to hold the scion/s in place and cover with grafting wax or sealing compound. This stops the cambium layers from drying out and also prevents the ingress of water into the cleft.

Whip

In whip grafting the scion and the stock are cut slanting and then joined. The grafted point is then bound with tape and covered with soft wax to prevent dehydration and germs. Also known as the whip and tongue graft, this is considered the most difficult to master but has the highest rate of success as it offers the most cambium contact between the two species. It is the most common graft used in top-dressing commercial fruit trees. It is generally used with stock less than $\frac{1}{2}$ in (1.3 cm) diameter, with the ideal diameter closer to $\frac{3}{8}$ in (0.95 cm) and the scion should be of roughly the same diameter as the stock.

Successful whip graft

Successful whip graft needing additional pruning the following season

The stock is cut through on one side only at a shallow angle with a sharp knife. (If the stock is a branch and not the main trunk of the rootstock then the cut surface should face outward from the centre of the tree.) The scion is similarly sliced through at an equal angle starting just below a bud, so that the bud is at the top of the cut and on the other side than the cut face.

A notch is cut downwards into the sliced face of the stock and a similar cut upwards into the face of the scion cut. These act as the tongues and it requires some skill to make the cuts so that the scion and the stock marry up neatly. The joint is then taped around and treated with tree-sealing compound or grafting wax.

The elongated "Z" shape adds strength, removing the need for a companion rod in the first season.

Stub

Stub grafting is a technique that requires less stock than cleft grafting, and retains the shape of a tree. Also scions are generally of 6-8 buds in this process.

Successful stub graft, healed

An incision is made into the branch 1 cm (0.39 in) long, then the scion is wedged and forced into the branch. The scion should be at an angle of at most 35° to the parent tree so that the crotch remains strong. The graft is covered with grafting compound.

After the graft has taken, the branch is removed and treated a few centimeters above the graft, to be fully removed when the graft is strong.

Four Flap (or Banana)

The four-flap graft (also called banana graft) is commonly used for pecans, and first became popular with this species in Oklahoma in 1975. It is heralded for maximum cambium overlap, but is a complex graft. It requires similarly sized diameters for the rootstock and graftwood. The bark of the rootstock is sliced and peeled back in four flaps, and the hardwood is removed, looking somewhat like a peeled banana. It is a difficult graft to learn.

Awl

Awl grafting takes the least resources and the least time. It is best done by an experienced grafter, as it is possible to accidentally drive the tool too far into the stock, reducing the scion's chance of survival. Awl grafting can be done by using a screwdriver to make a slit in the bark, not penetrating the cambium layer completely. Then inset the wedged scion into the incision.

Veneer

Veneer grafting, or inlay grafting, is a method used for stocks larger than three centimeters in diameter. The scion is recommended to be about as thick as a pencil. Clefts are made of the same size as the scion on the side of the branch, not on top. The scion end is shaped as a wedge, inserted, and wrapped with tape to the scaffolding branches to give it more strength.

Natural Grafting

Possible deliberate grafts on a Sessile Oak in Ayrshire, Scotland

A Husband and Wife tree - Natural grafting in blackthorn Prunus spinosa

Tree branches and more often roots of the same species will sometimes naturally graft; this is called inosculation. When roots make physical contact with each other they often grow together. A group of trees can share water and mineral nutrients via root grafts, which may be advantageous to weaker trees, and may also form a larger rootmass as an adaptation to promote fire resistance and regeneration as exemplified by the California black oak (*Quercus kelloggii*). Albino redwoods use root grafting as a form of plant parasitism of normal redwoods.

A problem with root grafts is that they allow transmission of certain pathogens, such as Dutch elm disease. Inosculation also sometimes occurs where two stems on the same tree, shrub or vine make contact with each other. This is common in plants such as strawberries and potato.

Graft Chimera

Occasionally, a so-called "graft hybrid" or more accurately graft chimera can occur where the tissues of the stock continue to grow within the scion. Such a plant can produce flowers and foliage typical of both plants as well as shoots intermediate between the two. The best-known example this is probably +*Laburnocytisus* 'Adamii', a graft hybrid between laburnum and broom, which originated in a nursery near Paris, France in 1825. This small tree bears yellow flowers typical of *Laburnum anagyroides*, purple flowers typical of *Chamaecytisus purpureus* and curious coppery-pink flowers that show characteristics of both "parents". Many species of cactus can also produce graft chimeras under the right conditions although they are often created unintentionally and such results are often hard to replicate.

Scientific Uses

Grafting has been important in flowering research. Leaves or shoots from plants induced to flower can be grafted onto uninduced plants and transmit a floral stimulus that induces them to flower.

The transmission of plant viruses has been studied using grafting. Virus indexing involves grafting a symptomless plant that is suspected of carrying a virus onto an indicator plant that is very susceptible to the virus.

Grafting can transfer chloroplasts (specialised DNA in plants that can conduct photosynthesis), mitichondrial DNA and the entire cell nucleus containing the genome to potentially make a new species making grafting a form of natural genetic engineering.

White Spruce

White spruce can be graffted with consistent success by using 8 cm to 10 cm scions of current growth on thrifty 4- to 5-year-old rootstock (Nienstaedt and Teich 1972). Before greenhouse grafting, rootstocks should be potted in late spring, allowed to make seasonal growth, then subjected to a period of chilling outdoors, or for about 8 weeks in a cool room at 2 °C (Nienstaedt 1966).

A method of grafting white spruce of seed-bearing age during the time of seed harvest in the fall was developed by Nienstaedt et al. (1958). Scions of white spruce of 2 ages of wood from 30- to 60-year-old trees were collected in the fall and grafted by 3 methods on potted stock to which different day-length treatments had been applied prior to grafting. The grafted stock were given long-day and natural-day treatments. Survival was 70% to 100% and showed effects of rootstock and post-grafting treatments in only a few cases. Photoperiod and temperature treatments after grafting, however, had considerable effect on scion activity and total growth. The best post-grafting treatment was 4 weeks of long-day treatment followed by 2 weeks of short-day treatment, then 8 weeks of chilling, and finally long-day treatment.

Since grafts of white spruce put on relatively little growth in the 2 years after grafting, techniques for accelerating the early growth were studied by Greenwood (1988) and others. The cultural regimes used to promote one additional growth cycle in one year involve manipulation of day length and the use of cold storage to satisfy chilling requirements. Greenwood took dormant potted grafts into the greenhouse in early January then gradually raised the temperature during the course of a week until the minimum temperature rose to 15 °C. Photoperiod was increased to 18 hours using incandescent lighting. In this technique, grafts are grown until elongation has been completed, normally by mid-March. Soluble 10-52-10 fertilizer is applied at both ends of the growth cycle and 20-20-20 during the cycle, with irrigation as needed. When growth elongation is complete, day length is reduced to 8 hours using a blackout curtain. Budset follows, and the grafts are held in the greenhouse until mid-May. Grafts are then moved into a cooler at 4 °C for 1000 hours, after which they are moved to a shade frame where they grow normally, with applications of fertilizer and irrigation as in the first cycle. Grafts are moved into cold frames or unheated greenhouse in September until January. Flower induction treatments are begun on grafts that have reached a minimum length of 1.0 m. Repotting from an initial pot size of 4.5litre to 16litre containers with a 2:1:1 soil mix of peat moss, loam, and aggregate.

In one of the first accelerated growth experiments, white spruce grafts made in January and February that would normally elongate shortly after grafting, set bud, and remain in that condition until the following spring, were refrigerated for 500, 1000, or 1500 hours beginning in mid-July, and a non-refrigerated control was held in the nursery. After completion of the cold treatment, the grafts were moved into the greenhouse with an 18-hour photoperiod until late October. Height increment was significantly (P 0.01) influenced by cold treatment. Best results were given by the 1000-hour treatment.

The refrigeration (cold treatment) phase was subsequently shown to be effective when applied 2 months earlier with proper handling and use of blackout curtains, which allows the second growth cycle to be completed in time to satisfy dormancy requirements before January (Greenwood et al. 1988).

Herbaceous Grafting

Grafting is often done for non -woody and vegetable plants (tomato, cucumber, eggplant and watermelon). Tomato grafting is very popular in Asia and Europe, and is gaining popularity in the United States. The main advantage of grafting is for disease-resistant rootstocks. Researchers in Japan developed automated processes using grafting robots as early as 1987. Plastic tubing can be used to prevent desiccation and support the healing at the graft/scion interface.

History

Grafting with detached scions has been practiced for thousands of years. It was in use by the Chinese before 2000 BC, then spread to the rest of Eurasia and was well established in ancient Greece.

Rootstock

A rootstock is part of a plant, often an underground part, from which new above-ground growth can be produced. It can refer to a rhizome or underground stem. In grafting, it refers to a plant, sometimes just a stump, which already has an established, healthy root system, onto which a cutting or a bud from another plant is grafted. In some cases, such as vines of grapes and other berries, cuttings may be used for rootstocks, the roots being established in nursery conditions before planting them out. The plant part grafted onto the rootstock is usually called the scion. The scion is the plant that has the properties that propagator desires above ground, including the photosynthetic activity and the fruit or decorative properties. The rootstock is selected for its interaction with the soil, providing the roots and the stem to support the new plant, obtaining the necessary soil water and minerals, and resisting the relevant pests and diseases. After a few weeks the tissues of the two parts will have grown together, eventually forming a single plant. After some years it may be difficult to detect the site of the graft although the product always contains the components of two genetically different plants.

The use of rootstocks is most commonly associated with fruiting plants and trees, and is useful for mass propagating many other types of plants that do not breed true from seed, or are particularly susceptible to disease when grown on their own roots.

Although grafting has been practiced for many hundreds of years, even in Roman times, most orchard rootstocks in current use were developed in the 20th century.

A variety of rootstocks may be used for a single species or cultivar of scion because different rootstocks impart different properties, such as vigour, fruit size and precocity. Rootstocks also may be selected for traits such as resistance to drought, root pests, and diseases. Grapevines for commercial planting are most often grafted onto rootstocks to avoid damage by phylloxera, though vines available for sale to back garden viticulturists may not be.

The rootstock may be a different species from the scion, but as a rule it should be closely related, For example, many commercial pears are grown on quince rootstock. Grafting can also be done in stages; a closely related scion is grafted to the rootstock, and a less closely related scion is grafted to the first scion. Serial grafting of several scions may also be used to produce a tree that bears several different fruit cultivars, with the same rootstock taking up and distributes water and minerals to the whole system. Those with more than three varieties are known as 'family trees'.

When it is difficult to match a plant to the soil in a certain field or orchard, growers may graft a scion onto a rootstock that is compatible with the soil. It may then be convenient to plant a range of ungrafted rootstocks to see which suit the growing conditions best; the fruiting characteristics of the scion may be considered later, once the most successful rootstock has been identified. Rootstocks are studied extensively and often are sold with a complete guide to their ideal soil and climate. Growers determine the pH, mineral content, nematode population, salinity, water availability, pathogen load and sandiness of their particular soil, and select a rootstock which is matched to it. Genetic testing is increasingly common, and new cultivars of rootstock are always being developed.

AxR1

AxR1 is a grape rootstock once widely used in California viticulture. Its name is an abbreviation for "Aramon Rupestris Ganzin No. 1", which in turn is based on its parentage: a cross (made by a French grape hybridizer named Ganzin) between Aramon, a *Vitis vinifera* cultivar, and Rupestris, an American grape species, *Vitis rupestris*—also used on its own as rootstock, "Rupestris St. George" or "St. George," referring to a town in the South of France, Saint Georges d'Orques, where it was popular.

It achieved a degree of notoriety in California when, after decades of recommendation as a preferred rootstock—despite repeated warnings from France and South Africa about its susceptibility (it had failed in Europe in the early 1900s), it ultimately succumbed to phylloxera in the 1980s, requiring the replanting of most of Napa and Sonoma, with disastrous financial consequences.

Most current day grape rootstocks were and are originally imported from Texas. These were taken from the native wild mustang grapes that grow across Texas. This rootstock also saved France's grape industry in the early 1900s when phylloxera decimated the wine and vineyards of Europe.

Fruit Tree Propagation

Fruit tree propagation is usually carried out vegetatively (non-sexually) by grafting or budding a desired variety onto a suitable rootstock.

Grafting, 1870, by Winslow Homer— an example of Grafting.

Perennial plants can be propagated either by sexual or vegetative means. Sexual reproduction begins when a male germ cell (pollen) from one flower fertilises a female germ cell (ovule, incipient seed) of the same species, initiating the development of a fruit containing seeds. Each seed, when germinated, can grow to become a new specimen tree. However, the new tree inherits characteristics of both its parents, and it will not grow "true" to the variety of either parent from which it came. That is, it will be a fresh individual with an unpredictable combination of characteristics of its own. Although this is desirable in terms of producing novel combinations from the richness of the gene pool of the two parent plants (such sexual recombination is the source of new cultivars), only rarely will the resulting new fruit tree be directly useful or attractive to the tastes of humankind. Most new plants will have characteristics that lie somewhere between those of the two parents.

Therefore, from the orchard grower or gardener's point of view, it is preferable to propagate fruit cultivars vegetatively in order to ensure reliability. This involves taking a cutting (or scion) of wood from a desirable parent tree which is then grown on to produce a new plant or "clone" of the original. In effect this means that the original Bramley apple tree, for example, was a successful variety grown from a pip, but that every Bramley since then has been propagated by taking cuttings of living matter from that tree, or one of its descendants.

Methods

The simplest method of propagating a tree vegetatively is rooting or taking cuttings. A cutting (usually a piece of stem of the parent plant) is cut off and stuck into soil. Artificial rooting hormones are sometimes used to improve chances of success. If the cutting does not die from rot-inducing fungi or desiccation first, roots grow from the buried portion of the cutting to become a new complete plant. However, although this works well for some plants (such as figs and olives), for most fruit tree cultivars this method has much too low a success rate to be commercially viable. Root cuttings (pieces of root cut off and induced to grow a new trunk) are also not used to propagate fruit trees, although this method is successful with some herbaceous plants.

A refinement on rooting is layering. This is rooting a piece of a wood that is still attached to its parent and continues to receive nourishment from it. The new plant is severed only after it has successfully grown roots. Layering is the technique most used for propagation of clonal apple rootstocks.

The most common method of propagating fruit trees, suitable for nearly all species, is grafting onto rootstocks. This in essence involves physically joining part of a shoot of a hybrid cultivar onto the roots of a different but closely related species or cultivar, so that the two parts grow together as one plant. The process of joining the two varieties must ensure maximum contact between the cambium (the layer just below the bark) of each, so that they grow together successfully. Grafting is a preferred method because it not only propagates a new plant of the desired hybrid cultivar, it usually also confers extra advantages as a result of the characteristics of the rootstocks (or stocks), which are selected for characteristics such as their vigour of growth, hardiness and soil tolerance, as well as compatibility with the desired variety that will form the aerial part of the plant (called the scion). For example, grape rootstocks descended from North American grapes allow European grapes to be grown in areas infested with *Phylloxera*, a soil-dwelling insect that attacks and kills European grapes when grown on their own roots. Two of the most common grafting techniques are "whip and tongue", carried out in spring as the sap rises, and "budding", which is performed around the end of summer.

Bud Grafting

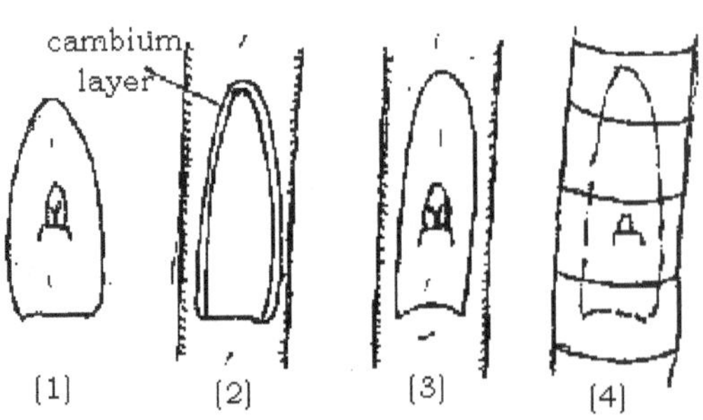

Diagram illustrating the bud grafting technique

1. Cut a slice of bud and bark from the parent tree.
2. Cut a similar sliver off the rootstock, making a little lip at the base to slot the scion into.
3. Join the two together and bind.
4. In time, the scion bud will grow into a shoot, which will develop into the desired tree.

Whip and Tongue Grafting

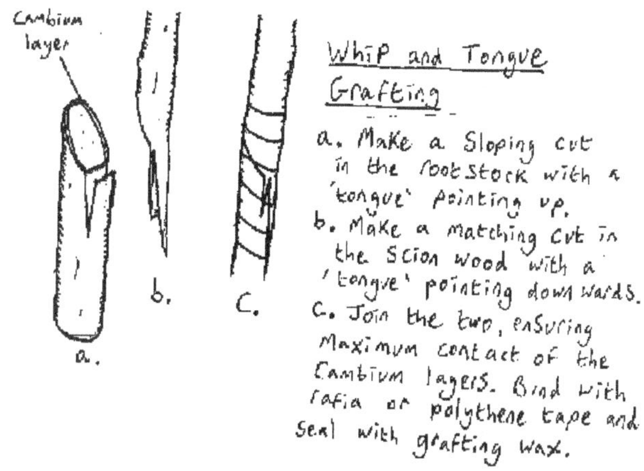

Diagram illustrating the whip and tongue grafting technique

1. Make a sloping cut in the rootstock with a "tongue" pointing upwards.
2. Make a matching cut in the scion wood with a "tongue" pointing downwards.
3. Join the two, ensuring maximum contact of the vascular cambium layers. Bind with raffia or polythene tape or wind around with a 5mm wide strip of elastic band (this is particularly successful because it keeps pressure on the cambium layers to be joined and eventually falls away without cutting into the bark as the tree grows) and seal with grafting wax.

Apple Rootstocks

One reason for grafting onto rootstocks is that this enables the grower to determine the tree's eventual size. Apple tree size classes number from one to ten in increasing height and breadth. A "1" is a dwarf which can be productive and as short as 3 feet (0.91 m) with proper pruning. A "10" is the standard sized tree with no dwarfing and will grow to 20 feet (6.1 m) tall and wide or more, dependent upon the variety chosen. In general the class range is (1) 10-20% of full size, (2) 20-30%, (3) 30-40% and so forth to size 10 which is 100% of full size.

The following are a selection of apple tree rootstocks. They are referred to by numbers prefixed by letters indicating the developer of the rootstock.

Very Small	Small	Medium	Large	Very Large
P22, M27, G65	G11, G41, M9, G16, Bud. 9, Mark	M26, G935, G202, G30, MM102, Interstems, M7, M116	MM106	MM111, Bud. 118, M25
6 ft/2m	8 ft/2.5m	10 ft/3m	14 ft/4m	18 ft/5m

”Bud 118” A winter hardy early bearing replacement for MM 111 bred in the Soviet Union. Full sized tree unless allowed to bear young which will stunt its growth. Hardy to USDA zone 3.

“Bud 9” A winter- hardy, early- bearing replacement for M9 bred in the Soviet Union. Dwarf tree resistant to crown rot and less susceptible to drought than most other dwarfing stocks. Produces large fruit, is precocious and hardy to USDA zone 3.

“M” designates Malling series developed stocks. East Malling Research is a pioneer in the development of dwarfing rootstocks. East Malling Research Station in Kent, England collected clones of the Paradise stocks from France in 1912 from which 24 “M” were designated with no particular order to the rootstock characteristics other than where they were located in the garden at the time the numbers were assigned. In other words, M.2 is a larger tree than M.9, while M.27 is smaller than M.26.

“MM” designates Malling-Merton stocks developed from joint breeding program by John Innes Institute, in Merton, England, & East Malling Research Station in the early 1950s. The “MM” series was developed primarily to provide resistance to Woolly Apple Aphid(Eriosomatinae) infestation.

“EMLA” designates East Malling / Long Ashton research stations who took the “M” stocks and developed virus free versions. For example., EMLA 7 is M 7 with a guaranteed virus-free stock. EMLA characteristics are often different from the parent “M” rootstock. Note that nearly all the apple rootstocks in the industry are now virus free.

“CG” or “G” designates Cornell-Geneva stocks which are those developed via the Cornell and USDA collaboration at the New York Agricultural Experiment Station in Geneva, NY. The “G” is the old designation. All newer stocks are “CG” followed by numbers that actually provide some information about the stock. As one might surmise, this is a huge improvement in the classical naming scheme which has no identification method at all.

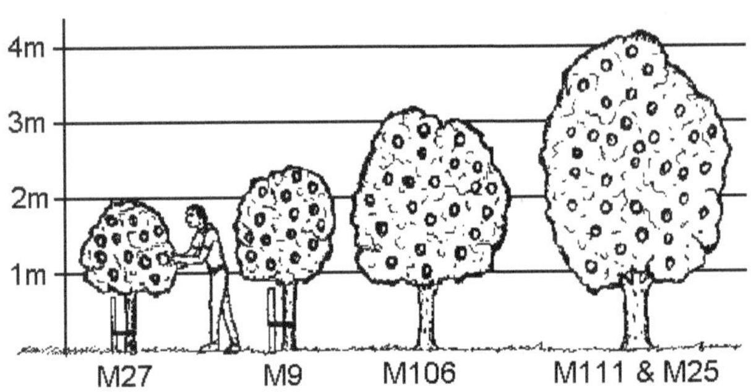

Diagram illustrating comparative sizes of apple trees depending on rootstock grafting

- M.27 Malling 27: A very dwarfing rootstock. Unless the central leader is supported, the tree will be very small. Often only used as an intermediate stem piece on MM.106 or MM.111. If handled and spaced properly, it can be a very productive stock for a vertical axe system. Trees can be grown three to four feet tall and produce about 45 fruit, roughly 2 pecks, depending on fruit cultivar.
- M.9: Very dwarfing - Reaches a height of 8 to 10 ft (2.4 to 3.0 m), coming into fruit after 3 to 4 years, reaching full capacity of 50 to 65 pounds (23 to 29 kg) after 5 to 6 years. It will grow under average soil conditions, but needs a good rich soil to thrive. A good choice where space is limited and fertility is high. Permanent staking is required, as is routine feeding

and watering. Trees on this rootstock always require leader support. The rootstock is very susceptible to fire blight and can develop burr knots.

- G.41 Geneva 41, released in 2005, produces trees the size of M.9. The rootstock was developed from a cross between M.27 and Robusta 5 made in 1975. Resistant to Crown|Collar|-Root rot(Phytophthora) and fire blight.
- M.26: Dwarfing - Similar to M9 in effect, although somewhat more vigorous and generally stronger, with a higher expected eventual yield of 65–75 pounds (29–34 kg) and height of 8 to 10 ft (2.4 to 3.0 m). A good choice where soil quality is average and compact growth is required. Comes into fruit after 3 to 4 years, reaching full cropping capacity after 5 to 6 years. Staking needed for first five years of its life. It is susceptible to collar rot and fire blight and should not be planted in a wet site. Certain varieties when grafted onto this rootstock may exhibit signs of graft union incompatibility, i.e., the union breaks.
- G.11 Geneva 11 is the second release of the Cornell breeding program similar in size to M.26 (Class 4), but more productive. Has the advantage of being resistant to fire blight and crown rot as well as only rarely producing suckers or burr knots.
- G.202 Geneva 202(CG 4202) is a semidwarfing rootstock that produces a tree in class 5 slightly larger than M.26(Size Class 4) and is more productive than M.26. It was developed from a cross of M.27 (Size Class 1) and Robusta 5 to be fire blight and Phytophthora resistant as well as having resistance to woolly apple aphids. In a 9-year study with the scion cultivar of the "Liberty" apple, G.202 was about 50 percent smaller than M.7, but had much greater production efficiency.
- M.7 Malling 7 rootstock produces a semidwarf tree of Class 6 that is freestanding in deep well drained soils but in rocky, steep, or shallow soils, it tends to lean. The rootstock may sucker profusely and is susceptible to collar rot(Phytophthora).
- MM.106: Semi-dwarfing — Sometimes referred to as semi-vigorous, this is the most widely used of rootstocks. It is probably the best choice for the average garden under average conditions, being tolerant of a wide range of soils, and producing a tree with an eventual size of 14 to 18 ft (4.3 to 5.5 m). Trees on this stock begin producing fruit within three to four years, and yield 90 to 110 pounds (41 to 50 kg) after some seven or eight years. MM106 is very suitable for use with weaker varieties that would produce undersized bushes with more dwarfing rootstocks. Can be trained as a half standard tree, but is rather too vigorous for cordons unless the soil is poor. Requires staking for the first five years or so of its life. Trees on MM.106 are highly susceptible to collar rot especially when planted in soils that remain wet(poor percolation).
- M.111 : Vigorous — Not generally suitable for garden scale growing, being both too large and spreading (18–25 ft), and too slow to come into cropping. They are, however, suitable for growing as specimen standards in the large garden, or for producing medium-sized bushes on poorer soils. Begins to fruit after six or seven years, reaching full capacity of 160 to 360 lb (73 to 163 kg) after eight to nine years. It is not winter hardy in United States zone 3 unless it receives abundant snow cover. Bud 118 was developed in the Soviet Union to replace MM 111, Bud 118 is winter hardy in zone 3 and very precocious. (Early bearing).

Planting depth of this rootstock is critical. The union should be no higher than 1 to 2 inches above the final soil line.

- M.25: Very vigorous — Suitable for a grassed orchard, and to grow on as a full standard. Plant 20 ft (6.1 m) apart, makes a tree of 15 to 20 ft (4.6 to 6.1 m) or more height and spread, eventually yielding 200 to 400 lb (91 to 181 kg) per tree. This rootstock is primarily used in UK and is rarely seen in the United States where M.111(size Class 8) is used for this size tree.
- Seedling: Very vigorous trees produced on a rootstock grown from seed. There is greater variability than with the vegetatively propagated rootstocks. Apples used for production of seedling rootstocks include "Dolgo" and "Antonovka", which are both extremely hardy and vigorous.

The Malling series and clones have been standard rootstocks for apples for many years and remain the "workhorses" for the commercial industry in the United States and the UK. However, since most of them are susceptible to disease some Malling rootstocks are being replaced by new breeds, including the Cornell-Geneva series, which has resistance to the major problems preventing quality production of apples utilizing organic control systems. One of the newest rootstocks of the "CG" series, only released commercially in 2004, is CG5202(G.202) which adds resistance to the woolly apple aphid, and when combined with highly resistant cultivars such as "Liberty" it is showing great potential.

Another desirable characteristic of rootstocks is environmental adaptability. This may be tolerance to wet/dry soil conditions, acidity/alkalinity of soil or even hot/cold air temperature. Rootstocks based on Siberian Crab apple are being used in colder areas for more cold tolerance.

The ability of new rootstocks to modify or augment characteristics of fruit trees is limited and may disappoint in the long term. It takes ten years to get a full picture of the effects of any one rootstock, so a rootstock that appears promising in the first five years of a trial may fail in the last five years. The Mark rootstock was such a stock and has now fallen mostly into disfavor. Another, the G.30, has proved to be an excellent stock for production but it was only after a number of years of trials that it was found to be somewhat incompatible with the cultivar "Gala", so that it is now recommended to be staked and wired.

An industry consortium undertakes trials of different rootstocks, called the "NC-140" trials. These test many pome rootstocks in many different sites across the United States and thereby provide growers a clearer picture of what to expect when growing fruit trees on specific stock, in specific planting methods in their specific area of the. This information has the potential to create economic benefits to both growers and consumers, as well as helping to reduce the need to spray pesticides as frequently as is currently required.

Pear Rootstocks

Pears are usually grafted onto quince rootstocks, which produce small to medium-sized trees. Some varieties, however, are not compatible with quince, and these require double working. This means that a piece of pear graft-work compatible with both the quince rootstock and the pear variety is used as an intermediate between the two. If this is not done the pear and the rootstock could eventually separate at the graft. Varieties that require double working include "Bristol Cross", "Dr Jules Guyot", "Doyenné d' été" and "Williams Bon Chrétien".

- Quince C: Moderately vigorous — Makes a bush pear tree about 8 to 18 ft (2.4 to 5.5 m) tall, bearing fruit within four to eight years. Suitable for highly fertile soils and vigorous varieties, but not where conditions are poor. Used for bush, cordon and espalier growing. Old stocks of Quince C may be infected with a virus, so care should be taken to obtain certified virus free stock. If in doubt, use Quince A as there is not a great amount of difference in vigor between the two.
- Quince A: Medium vigor — Slightly more vigorous than Quince C, this is the most common variety upon which pears are grafted. Bears fruit between four and eight years, making a tree of some 10 to 20 ft (3.0 to 6.1 m) in height and spread. Suitable for all forms of pear trees except standards.

Pear stock: Very vigorous — Pears grafted onto pear rootstocks make very large standard trees, not suitable for most gardens.

Cherries

Until the 1970s, cherries were grown on the vigorous Malling F12/1, Mazzard (*Prunus avium*), or Maheleb (*P. maheleb*) rootstocks, which required much space and time before cropping began, thus the growing of cherries was not a realistic option on a garden scale. The introduction of the rootstock "Colt" enabled trees reaching a maximum height of 12 to 15 ft (3.7 to 4.6 m) to be grown, and if trained as a pyramid it is possible to restrict growth to about 10 ft (3.0 m). The popular sweet variety "Stella" can even be successfully grown in a pot on the patio when grafted onto a "Colt" rootstock. A newer rootstock, Gisela 5, is slowly becoming available to gardeners and produces a tree 20% smaller than Colt and 45% smaller than Mahaleb and Mazzard, making netting for bird protection much easier. Furthermore, German Nurseries Consortium (Consortium Deutscher Baumschulen — CDB®) is introducing across the EU their newest most dwarfing Gisela 3 cherry rootstock that has 50% dwarfing qualities in comparison with Mahaleb and Mazzard and is 10% smaller still than Gisela 5 rootstock.

Plums

Plum rootstocks include:

- Pixy — A dwarfing rootstock (*Prunus domestica* subsp. *insititia*, or sometimes *Prunus insititia*) suitable for bush trees planted 8 to 10 ft (2.4 to 3.0 m) apart.
- St. Julien A — A semi-vigorous rootstock suitable for bush and half standards planted 12 to 15 ft (3.7 to 4.6 m) apart. Also suitable for peaches, nectarines and apricots.
- Brompton or Myrobalan B- (Prunus cerasifera) — Suitable for half standards planted 18 to 22 ft (5.5 to 6.7 m) apart. Also suitable for peaches, nectarines and apricots.
- Myro-29C — (Prunus cerasifera) Semi-dwarf rootstock. Shallow, vigorous, good choice for hard soils. Somewhat drought tolerant.
- Citation — Semi-dwarf rootstock. Shallow, vigorous, good choice for hard soils. Prefers a wetter soil.

Own-root Fruit Trees

Many species of fruit, e.g., fig, filbert, olive, and pomegranate, are commonly grown on their own roots, as there may be no great advantages to using a special rootstock, or suitable rootstocks may not be readily available. However, even for fruit trees that usually are grown grafted on a rootstock, there can be advantages in growing them on their own roots instead, particularly in the traditional coppicing systems advocated in both sustainable agriculture and permaculture. Disadvantages of using own-root trees can include excessive size and excessive production of wood (thus very long times until the start of fruit production), although training branches horizontally and limiting pruning to summer only may help encourage fruit production at an earlier age. There is a lack of research on the use of the own root method in large-scale systems. British horticulturalist Hugh Ermen was a prominent advocate of own root fruit trees, particularly Apple trees, following his work at the Brogdale research centre in Kent. The research undertaken at Brogdale did not continue but there has been renewed academic interest in own root fruit trees since the turn of the century.

Family Trees and Fruit Salad Trees

In addition to propagating trees on rootstocks designed to control size/vigour and confer disease resistance, grafting above the rootstock can be used to provide multiple cultivars of a single species, known as a family tree, or, within certain limits, cultivars of different fruit species on one tree, often known as a fruit salad tree. Family trees typically combine several cultivars (two or three being most common) of apple, pear or a given species of stonefruit on a single rootstock, while fruit salad trees typically carry two or more different species from within a given genus, such as plum, apricot, and peach or mandarin orange, lemon, and lime. Certain combinations, including sour cherry (*prunus cerasus*)/sweet cherry (*prunus avium*), although from the same genus, are known to be difficult, although successes have sometimes been reported. Other grafts of this kind can produce the Pomato.

Intercropping

Intercropping is a multiple cropping practice involving growing two or more crops in proximity. The most common goal of intercropping is to produce a greater yield on a given piece of land by making use of resources that would otherwise not be utilized by a single crop. Careful planning is required, taking into account the soil, climate, crops, and varieties. It is particularly important not to have crops competing with each other for physical space, nutrients, water, or sunlight. Examples of intercropping strategies are planting a deep-rooted crop with a shallow-rooted crop, or planting a tall crop with a shorter crop that requires partial shade. Inga alley cropping has been proposed as an alternative to the ecological destruction of slash-and-burn farming.

When crops are carefully selected, other agronomic benefits are also achieved. Lodging-prone plants, those that are prone to tip over in wind or heavy rain, may be given structural support by their companion crop. Creepers can also benefit from structural support. Some plants are used to suppress weeds or provide nutrients. Delicate or light-sensitive plants may be given shade or protection, or otherwise wasted space can be utilized. An example is the tropical multi-tier system where coconut occupies the upper tier, banana the middle tier, and pineapple, ginger, or leguminous fodder, medicinal or aromatic plants occupy the lowest tier.

Intercropping of compatible plants also encourages biodiversity, by providing a habitat for a variety of insects and soil organisms that would not be present in a single-crop environment. This in turn can help limit outbreaks of crop pests by increasing predator biodiversity. Additionally, reducing the homogeneity of the crop increases the barriers against biological dispersal of pest organisms through the crop.

The degree of spatial and temporal overlap in the two crops can vary somewhat, but both requirements must be met for a cropping system to be an intercrop. Numerous types of intercropping, all of which vary the temporal and spatial mixture to some degree, have been identified. These are some of the more significant types:

- Mixed intercropping, as the name implies, is the most basic form in which the component crops are totally mixed in the available space.
- Row cropping involves the component crops arranged in alternate rows. Variations include alley cropping, where crops are grown in between rows of trees, and strip cropping, where multiple rows, or a strip, of one crop are alternated with multiple rows of another crop. A new version of this is to intercrop rows of solar photovoltaic modules with agriculture crops. This practice is called agrivoltaics.
- Temporal intercropping uses the practice of sowing a fast-growing crop with a slow-growing crop, so that the fast-growing crop is harvested before the slow-growing crop starts to mature.
- Further temporal separation is found in relay cropping, where the second crop is sown during the growth, often near the onset of reproductive development or fruiting, of the first crop, so that the first crop is harvested to make room for the full development of the second.

Chili pepper intercropped with coffee in Colombia's southwestern Cauca Department

Coconut and Tagetes erecta, a multilayer cropping in India

Horticultural Oil

Horticultural Oils or Narrow Range Oils are lightweight oils, either petroleum or vegetable based. They are used in both horticulture and agriculture, where they are applied as a dilute spray on plant surfaces to control insects and mites. They are also sometimes included in tank mixes as a surfactant.

The oils provide control by smothering the target pests, and are only effective if applied directly to the pest, and provide no residual controls.

Oils are generally considered suitable for 'organic pest control', with most oils permitted under the U.S. National Organic Program.

Terms Used to Describe Oils

Dormant oil: An oil used on woody plants during the dormant season. This term originally referred to heavier weight, less well-refined oils that were unsafe to use on plants after they broke dormancy. However, these older oils have been replaced with more refined, light-weight oils that have potential application to plant foliage. Dormant oil now refers to the time of application rather than to any characteristic type of oil.

Horticultural oils: An oil used to control a pest on plants.

Mineral oil: A petroleum-derived oil (as opposed to vegetable oils).

Narrow-range oil: A highly refined oil that has a narrow range of distillation. Narrow-range oils fall in the superior oil classification. The terms may be used nearly interchangeably.

Spray oil: An oil designed to be mixed with water and applied to plants as a spray for pest control.

Summer oil: An oil used on plants when foliage is present (also called foliar oils). As with dormant oil, the term now refers to the time an application is made rather than to the properties of the oil.

Supreme oil: A term used to categorize highly refined oils that distill at slightly higher temperatures and over a wider range than the narrow-range oils. Most supreme oils meet the characteristics of a superior oil.

Superior oil: A term originated by P.J. Chapman in 1947 to categorize summer-use oils that met certain specifications. This included a high proportion of paraffinic hydrocarbons and purification that allowed year-round use without phytotoxicity. Since then, further developments have resulted in oils that distill over a narrow temperature range. Most superior oils are now better referred to as narrow-range oils.

Vegetable oil: An oil derived from the seeds of some oil seed crop (e.g., soybeans, canola, cottonseed).

Gardening

Gardening is the practice of growing and cultivating plants as part of horticulture. In gardens, ornamental plants are often grown for their flowers, foliage, or overall appearance; useful plants, such as root vegetables, leaf vegetables, fruits, and herbs, are grown for consumption, for use as dyes, or for medicinal or cosmetic use. Gardening is considered to be a relaxing activity for many people.

Part of a parterre in an English garden

Gardening ranges in scale from fruit orchards, to long boulevard plantings with one or more different types of shrubs, trees, and herbaceous plants, to residential yards including lawns and foundation plantings, to plants in large or small containers grown inside or outside. Gardening may be very specialized, with only one type of plant grown, or involve a large number of different plants in mixed plantings. It involves an active participation in the growing of plants, and tends to be labor-intensive, which differentiates it from farming or forestry.

History

Robert Hart's forest garden in Shropshire, England

Ancient Times

Forest gardening, a forest-based food production system, is the world's oldest form of gardening. Forest gardens originated in prehistoric times along jungle-clad river banks and in the wet foothills of monsoon regions. In the gradual process of families improving their immediate environment, useful tree and vine species were identified, protected and improved while undesirable species were eliminated. Eventually foreign species were also selected and incorporated into the gardens.

After the emergence of the first civilizations, wealthy individuals began to create gardens for aesthetic purposes. Egyptian tomb paintings from around 1500 BC provide some of the earliest physical evidence of ornamental horticulture and landscape design; they depict lotus ponds surrounded by symmetrical rows of acacias and palms. A notable example of ancient ornamental gardens were the Hanging Gardens of Babylon—one of the Seven Wonders of the Ancient World —while ancient Rome had dozens of gardens.

Wealthy ancient Egyptians used gardens for providing shade. Egyptians associated trees and gardens with gods as they believed that their deities were pleased by gardens. Gardens in ancient Egypt were often surrounded by walls with trees planted in rows. Among the most popular species planted were date palms, sycamores, fir trees, nut trees, and willows. These gardens were a sign of higher socioeconomic status. In addition, wealthy ancient Egyptians grew vineyards, as wine was a sign of the higher social classes. Roses, poppies, daisies and irises could all also be found in the gardens of the Egyptians.

Assyria was also renowned for its beautiful gardens. These tended to be wide and large, some of them used for hunting game—rather like a game reserve today—and others as leisure gardens. Cypresses and palms were some of the most frequently planted types of trees.

Ancient Roman gardens were laid out with hedges and vines and contained a wide variety of flowers—acanthus, cornflowers, crocus, cyclamen, hyacinth, iris, ivy, lavender, lilies, myrtle, narcissus, poppy, rosemary and violets—as well as statues and sculptures. Flower beds were popular in the courtyards of rich Romans.

A gardener at work, 1607

The Middle Ages

The Middle Age represented a period of decline in gardens for aesthetic purposes, with regard to gardening. After the fall of Rome, gardening was done for the purpose of growing medicinal herbs and/or decorating church altars. Monasteries carried on a tradition of garden design and intense horticultural techniques during the medieval period in Europe. Generally, monastic garden types consisted of kitchen gardens, infirmary gardens, cemetery orchards, cloister garths and vineyards. Individual monasteries might also have had a "green court", a plot of grass and trees where horses could graze, as well as a cellarer's garden or private gardens for obedientiaries, monks who held specific posts within the monastery.

Islamic gardens were built after the model of Persian gardens and they were usually enclosed by walls and divided in 4 by watercourses. Commonly, the center of the garden would have a pool or pavilion. Specific to the Islamic gardens are the mosaics and glazed tiles used to decorate the rills and fountains that were built in these gardens.

By the late 13th century, rich Europeans began to grow gardens for leisure and for medicinal herbs and vegetables. They surrounded the gardens by walls to protect them from animals and to provide seclusion. During the next two centuries, Europeans started planting lawns and raising flowerbeds and trellises of roses. Fruit trees were common in these gardens and also in some, there were turf seats. At the same time, the gardens in the monasteries were a place to grow flowers and medicinal herbs but they were also a space where the monks could enjoy nature and relax.

The gardens in the 16th and 17th century were symmetric, proportioned and balanced with a more classical appearance. Most of these gardens were built around a central axis and they were divided into different parts by hedges. Commonly, gardens had flowerbeds laid out in squares and separated by gravel paths.

Gardens in Renaissance were adorned with sculptures, topiary and fountains. In the 17th century, knot gardens became popular along with the hedge mazes. By this time, Europeans started planting new flowers such as tulips, marigolds and sunflowers.

Cottage Gardens

Cottage gardens, which emerged in Elizabethan times, appear to have originated as a local source for herbs and fruits. One theory is that they arose out of the Black Death of the 1340s, when the death of so many laborers made land available for small cottages with personal gardens. According to the late 19th-century legend of origin, these gardens were originally created by the workers that lived in the cottages of the villages, to provide them with food and herbs, with flowers planted among them for decoration. Farm workers were provided with cottages that had architectural quality set in a small garden—about an acre—where they could grow food and keep pigs and chickens.

Authentic gardens of the yeoman cottager would have included a beehive and livestock, and frequently a pig and sty, along with a well. The peasant cottager of medieval times was more interested in meat than flowers, with herbs grown for medicinal use rather than for their beauty. By Elizabethan times there was more prosperity, and thus more room to grow flowers. Even the early cottage garden flowers typically had their practical use—violets were spread on the floor (for their pleasant scent and keeping out vermin); calendulas and primroses were both attractive and used in cooking. Others, such as sweet william and hollyhocks, were grown entirely for their beauty.

18th Century

In the 18th century, gardens were laid out more naturally, without any walls. This style of smooth undulating grass, which would run straight to the house, clumps, belts and scattering of trees and his serpentine lakes formed by invisibly damming small rivers, were a new style within the English landscape, a "gardenless" form of landscape gardening, which swept away almost all the remnants of previous formally patterned styles. The English garden usually included a lake, lawns set against groves of trees, and often contained shrubberies, grottoes, pavilions, bridges and follies such as mock temples, Gothic ruins, bridges, and other picturesque architecture, designed to recreate an idyllic pastoral landscape. This new style emerged in England in the early 18th century, and spread across Europe, replacing the more formal, symmetrical Garden à la française of the 17th century as the principal gardening style of Europe. The English garden presented an idealized view of nature. They were often inspired by paintings of landscapes by Claude Lorraine and Nicolas Poussin, and some were Influenced by the classic Chinese gardens of the East, which had recently been described by European travelers. The work of Lancelot 'Capability' Brown was particularly influential. Also, in 1804 the Horticultural Society was formed. Gardens of the 19th century contained plants such as the monkey puzzle or Chile pine. This is also the time when the so-called "gardenesque" style of gardens evolved. These gardens displayed a wide variety of flowers in a rather small space. Rock gardens increased in popularity in the 19th century.

Types

Conservatory of Flowers in Golden Gate Park, San Francisco

Hanging baskets in Thornbury, South Gloucestershire

Residential gardening takes place near the home, in a space referred to as the garden. Although a garden typically is located on the land near a residence, it may also be located on a roof, in an atrium, on a balcony, in a windowbox, or on a patio or vivarium.

Gardening also takes place in non-residential green areas, such as parks, public or semi-public gardens (botanical gardens or zoological gardens), amusement and amusement parks, along transportation corridors, and around tourist attractions and garden hotels. In these situations, a staff of gardeners or groundskeepers maintains the gardens.

- Indoor gardening is concerned with the growing of houseplants within a residence or building, in a conservatory, or in a greenhouse. Indoor gardens are sometimes incorporated as part of air conditioning or heating systems. Indoor gardening extends the growing season in the fall and spring and can be used for winter gardening.
- Native plant gardening is concerned with the use of native plants with or without the intent of creating wildlife habitat. The goal is to create a garden in harmony with, and adapted to a given area. This type of gardening typically reduces water usage, maintenance, and fertilization costs, while increasing native faunal interest.
- Water gardening is concerned with growing plants adapted to pools and ponds. Bog gardens are also considered a type of water garden. These all require special conditions and considerations. A simple water garden may consist solely of a tub containing the water and plant(s). In aquascaping, a garden is created within an aquarium tank.
- Container gardening is concerned with growing plants in any type of container either indoors or outdoors. Common containers are pots, hanging baskets, and planters. Container gardening is usually used in atriums and on balconies, patios, and roof tops.
- Hügelkultur is concerned with growing plants on piles of rotting wood, as a form of raised bed gardening and composting in situ. An English loanword from German, it means "mound garden." Toby Hemenway, noted Permaculture author and teacher, considers wood buried in trenches to also be a form of hugelkultur referred to as a dead wood swale. Hugelkultur is practiced by Sepp Holzer as a method of forest gardening and agroforestry, and by Geoff Lawton as a method of dryland farming and desert greening. When used as a method of disposing of large volumes of waste wood and woody debris, hugelkultur accomplishes carbon sequestration. It is also a form of xeriscaping.
- Community gardening is a social activity in which an area of land is gardened by a group of people, providing access to fresh produce and plants as well as access to satisfying labor, neighborhood improvement, sense of community and connection to the environment. Community gardens are typically owned in trust by local governments or nonprofits.
- Garden sharing partners landowners with gardeners in need of land. These shared gardens, typically front or back yards, are usually used to produce food that is divided between the two parties.
- Organic gardening uses natural, sustainable methods, fertilizers and pesticides to grow non-genetically modified crops.

Garden Features and Accessories

There is a wide range of features and accessories available in the market for both the professional gardener and the amateur to exercise their creativity. These are used to add decoration or functionality, and may be made from a wide range of materials such as copper, stone, wood, bamboo, stainless steel, clay, stained glass, concrete, or iron. Examples include trellis, arbors, statues, benches, water fountains, urns, bird baths and feeders, and garden lighting such as candle lanterns and oil lamps. The use of these items can be part of the expression of a gardener's gardening personality.

Gardening Departments and Centers

Gardening departments and centers mainly sell plants, sundries, and garden accessories, but in recent times, many now stock outdoor leisure products as diverse as spas, furniture, and barbecues. Many garden centers now include food halls, and sections for clothing, gifts, pets, and power tools. There are also a number of online garden centers that now deliver directly to customers' doors.

Comparison with Farming

Gardening for beauty is likely nearly as old as farming for food, however for most of history for the majority of people there was no real distinction since the need for food and other useful product trumped other concerns. Small-scale, subsistence agriculture (called hoe-farming) is largely indistinguishable from gardening. A patch of potatoes grown by a Peruvian peasant or an Irish smallholder for personal use could be described as either a garden or a farm. Gardening for average people evolved as a separate discipline, more concerned with aesthetics and recreation, under the influence of the pleasure gardens of the wealthy. Meanwhile, farming has evolved (in developed countries) in the direction of commercialization, economics of scale, and monocropping.

Hand gardening tools

In respect to its food producing purpose, gardening is distinguished from farming chiefly by scale and intent. Farming occurs on a larger scale, and with the production of salable goods as a major motivation. Gardening is done on a smaller scale, primarily for pleasure and to produce goods for the gardener's own family or community. There is some overlap between the terms, particularly in that some moderate-sized vegetable growing concerns, often called market gardening, can fit in either category.

Planting in a garden

The key distinction between gardening and farming is essentially one of scale; gardening can be a hobby or an income supplement, but farming is generally understood as a full-time or commercial activity, usually involving more land and quite different practices. One distinction is that gardening is labor-intensive and employs very little infrastructural capital, sometimes no more than a few tools, e.g. a spade, hoe, basket and watering can. By contrast, larger-scale farming often involves irrigation systems, chemical fertilizers and harvesters or at least ladders, e.g. to reach up into fruit trees. However, this distinction is becoming blurred with the increasing use of power tools in even small gardens.

In part because of labor intensity and aesthetic motivations, gardening is very often much more productive per unit of land than farming. In the Soviet Union, half the food supply came from small peasants' garden plots on the huge government-run collective farms, although they were tiny patches of land. Some argue this as evidence of the superiority of capitalism, since the peasants were generally able to sell their produce. Others consider it to be evidence of a tragedy of the commons, since the large collective plots were often neglected, or fertilizers or water redirected to the private gardens.

The term precision agriculture is sometimes used to describe gardening using intermediate technology (more than tools, less than harvesters), especially of organic varieties. Gardening is effectively scaled up to feed entire villages of over 100 people from specialized plots. A variant is the community garden which offers plots to urban dwellers.

Gardens as Art

Garden design is considered to be an art in most cultures, distinguished from gardening, which generally means *garden maintenance*. Garden design can include different themes such as perennial, butterfly, wildlife, Japanese, water, tropical, or shade gardens. In Japan, Samurai and Zen monks were often required to build decorative gardens or practice related skills like flower arrangement known as *ikebana*. In 18th-century Europe, country estates were refashioned by landscape gardeners into formal gardens or landscaped park lands, such as at Versailles, France, or Stowe, England. Today, landscape architects and garden designers continue to produce artistically creative designs for private garden spaces. In the USA, professional landscape designers are certified by the Association of Professional Landscape Designers.

Garden at the Schultenhof in Mettingen, North Rhine-Westphalia, Germany

Social Aspects

People can express their political or social views in gardens, intentionally or not. The lawn vs. garden issue is played out in urban planning as the debate over the "land ethic" that is to determine urban land use and whether hyper hygienist bylaws (e.g. weed control) should apply, or whether land should generally be allowed to exist in its natural wild state. In a famous Canadian Charter of Rights case, "Sandra Bell vs. City of Toronto", 1997, the right to cultivate all native species, even most varieties deemed noxious or allergenic, was upheld as part of the right of free expression.

Community gardening comprises a wide variety of approaches to sharing land and gardens.

People often surround their house and garden with a hedge. Common hedge plants are privet, hawthorn, beech, yew, leyland cypress, hemlock, arborvitae, barberry, box, holly, oleander, forsythia and lavender. The idea of open gardens without hedges may be distasteful to those who enjoy privacy. The Slow Food movement has sought in some countries to add an edible school yard and garden classrooms to schools, e.g. in Fergus, Ontario, where these were added to a public school to augment the kitchen classroom. Garden sharing, where urban landowners allow gardeners to grow on their property in exchange for a share of the harvest, is associated with the desire to control the quality of one's food, and reconnect with soil and community.

In US and British usage, the production of ornamental plantings around buildings is called *landscaping*, *landscape maintenance* or *grounds keeping*, while international usage uses the term *gardening* for these same activities.

Also gaining popularity is the concept of "Green Gardening" which involves growing plants using organic fertilizers and pesticides so that the gardening process - or the flowers and fruits produced thereby - doesn't adversely affect the environment or people's health in any manner.

Garden Pests

Garden pests are generally plants, fungi, or animals (frequently insects) that engages in activity that the gardener considers undesirable. It may crowd out desirable plants, disturb soil, stunt the growth of young seedlings, steal or damage fruit, or otherwise kill plants, hamper their growth, damage their appearance, or reduce the quality of the edible or ornamental portions of the plant. Aphids, spider mites, slugs, snails, ants, birds, and even cats are commonly considered to be garden pests.

Because gardeners may have different goals, organisms considered "garden pests" vary from gardener to gardener. *Tropaeolum speciosum*, for example, may be considered a desirable and ornamental garden plant, or it may be considered a pest if it seeds and starts to grow where it is not wanted. As another example, in lawns, moss can become dominant and be impossible to eradicate. In some lawns, lichens, especially very damp lawn lichens such as *Peltigera lactucfolia* and *P. membranacea*, can become difficult to control and be considered pests.

Garden Pest Control

There are many ways by which unwanted pests are removed from a garden. The techniques vary depending on the pest, the gardener's goals, and the gardener's philosophy. For example, snails may be dealt with through the use of a chemical pesticide, an organic pesticide, hand-picking, barriers, or simply growing snail-resistant plants.

Pest control is often done through the use of pesticides, which may be either organic or artificially synthesized. Pesticides may affect the ecology of a garden due to their effects on the populations of both target and non-target species. For example, unintended exposure to some neonicotinoid pesticides has been proposed as a factor in the recent decline in honey bee populations. A mole vibrator can deter mole activity in a garden.

Other means of control include the removal of infected plants, using fertilizers and biostimulants to improve the health and vigour of plants so they better resist attack, practising crop rotation to prevent pest build-up, using companion planting, and practising good garden hygiene, such as disinfecting tools and clearing debris and weeds which may harbour pests.

Controlled-environment Agriculture

Controlled-environment agriculture (CEA) is a technology-based approach toward food production. The aim of CEA is to provide protection and maintain optimal growing conditions throughout the development of the crop. Production takes place within an enclosed growing structure such as a greenhouse or building. Plants are often grown using hydroponic methods in order to supply the proper amounts of water and nutrients to the root zone. CEA optimizes the use of resources such as water, energy, space, capital and labor. CEA technologies include hydroponics, aquaculture, and aquaponics.

Controllable variables:

- Temperature (air, nutrient solution, root-zone)
- Humidity (%RH)
- Carbon dioxide (CO_2)

- Light (intensity, spectrum, interval)
- Nutrient concentration (PPM, EC)
- Nutrient pH (acidity)

CEA facilities can range from fully automated glasshouses with computer controls for watering, lighting and ventilation, to low-tech solutions such as cloches or plastic film on field grown crops and plastic-covered tunnels.

CEA is used in research so that a specific aspect of production can be isolated while all other variables remain the same. Tinted glass could be compared to plain glass in this way during an investigation into photosynthesis. Another possibility would be an investigation into the use of supplementary lighting for growing lettuce under a hydroponic system.

A February 2011 article in the magazine *Science Illustrated* states, "In commercial agriculture, CEA can increase efficiency, reduce pests and diseases, and save resources. ... Replicating a conventional farm with computers and LED lights is expensive but proves cost-efficient in the long run by producing up to 20 times as much high-end, pesticidee-free produce as a similar-size plot of soil. Fourteen thousand square feet of closely monitored plants produce 15 million seedlings annually at the solar-powered factory. Such factories will be necessary to meet urban China's rising demand for quality fruits and vegetables."

Hydroponics

Hydroponics is a subset of hydroculture and is a method of growing plants using mineral nutrient solutions, in water, without soil. Terrestrial plants may be grown with their roots in the mineral solution only, or in an inert medium, such as perlite or gravel. The nutrients in hydroponics can be from fish waste, normal nutrients, or duck manure.

NASA researcher checking hydroponic onions with Bibb lettuce to his left and radishes to the right

History

The earliest published work on growing terrestrial plants without soil was the 1627 book *Sylva Sylvarum* by Francis Bacon, printed a year after his death. Water culture became a popular research technique after that. In 1699, John Woodward published his water culture experiments with spearmint. He found that plants in less-pure water sources grew better than plants in distilled water. By 1842, a list of nine elements believed to be essential for plant growth had been compiled, and the discoveries of German botanists Julius von Sachs and Wilhelm Knop, in the years 1859–1875, resulted in a development of the technique of soilless cultivation. Growth of terrestrial plants without soil in mineral nutrient solutions was called solution culture. It quickly became a standard research and teaching technique and is still widely used. Solution culture is now considered a type of hydroponics where there is no inert medium.

In 1929, William Frederick Gericke of the University of California at Berkeley began publicly promoting that solution culture be used for agricultural crop production. He first termed it aquaculture but later found that aquaculture was already applied to culture of aquatic organisms. Gericke created a sensation by growing tomato vines twenty-five feet high in his back yard in mineral nutrient solutions rather than soil. He introduced the term hydroponics, water culture, in 1937, proposed to him by W. A. Setchell, a phycologist with an extensive education in the classics. Hydroponics is derived from neologism onstructed in analogy to geoponica, that which concerns agriculture, replacing, earth, with water.

Reports of Gericke's work and his claims that hydroponics would revolutionize plant agriculture prompted a huge number of requests for further information. Gericke had been denied use of the University's greenhouses for his experiments due to the administration's skepticism, and when the University tried to compel him to release his preliminary nutrient recipes developed at home he requested greenhouse space and time to improve them using appropriate research facilities. While he was eventually provided greenhouse space, the University assigned Hoagland and Arnon to re-develop Gericke's formula and show it held no benefit over soil grown plant yields, a view held by Hoagland. In 1940, Gericke published the book, *Complete Guide to Soil less Gardening,* after leaving his academic position in a climate that was politically unfavorable.

Two other plant nutritionists at the University of California were asked to research Gericke's claims. Dennis R. Hoagland and Daniel I. Arnon wrote a classic 1938 agricultural bulletin, *The Water Culture Method for Growing Plants Without Soil,*. Hoagland and Arnon claimed that hydroponic crop yields were no better than crop yields with good-quality soils. Crop yields were ultimately limited by factors other than mineral nutrients, especially light. This research, however, overlooked the fact that hydroponics has other advantages including the fact that the roots of the plant have constant access to oxygen and that the plants have access to as much or as little water as they need. This is important as one of the most common errors when growing is over- and under- watering; and hydroponics prevents this from occurring as large amounts of water can be made available to the plant and any water not used, drained away, recirculated, or actively aerated, eliminating anoxic conditions, which drown root systems in soil. In soil, a grower needs to be very experienced to know exactly how much water to feed the plant. Too much and the plant will be unable to access oxygen; too little and the plant will lose the ability to transport nutrients, which are typically moved into the roots while in solution. These two researchers developed several formulas for mineral nutrient solutions, known as Hoagland solution. Modified Hoagland solutions are still in use.

One of the earliest successes of hydroponics occurred on Wake Island, a rocky atoll in the Pacific Ocean used as a refuelling stop for Pan American Airlines. Hydroponics was used there in the 1930s to grow vegetables for the passengers. Hydroponics was a necessity on Wake Island because there was no soil, and it was prohibitively expensive to airlift in fresh vegetables.

In the 1960s, Allen Cooper of England developed the Nutrient film technique. The Land Pavilion at Walt Disney World's EPCOT Center opened in 1982 and prominently features a variety of hydroponic techniques. In recent decades, NASA has done extensive hydroponic research for its Controlled Ecological Life Support System (CELSS). Hydroponics intended to take place on Mars are using LED lighting to grow in a different color spectrum with much less heat.

Origin

Soilless Culture

Gericke originally defined hydroponics as crop growth in mineral nutrient solutions. Hydroponics is a subset of soilless culture. Many types of soilless culture do not use the mineral nutrient solutions required for hydroponics.

Plants that are not traditionally grown in a climate would be possible to grow using a controlled environment system like hydroponics. NASA has also looked to utilize hydroponics in the space program. Ray Wheeler, a plant physiologist at Kennedy Space Center's Space Life Science Lab, believes that hydroponics will create advances within space travel. He terms this as a bioregenerative life support system.

Techniques

There are two main variations for each medium, sub-irrigation and top irrigation. For all techniques, most hydroponic reservoirs are now built of plastic, but other materials have been used including concrete, glass, metal, vegetable solids, and wood. The containers should exclude light to prevent algae growth in the nutrient solution.

Static Solution Culture

The deep water raft tank at the CDC South Aquaponics greenhouse in Brooks, Alberta.

In static solution culture, plants are grown in containers of nutrient solution, such as glass Mason jars (typically, in-home applications), plastic buckets, tubs, or tanks. The solution is usually gently aerated but may be un-aerated. If un-aerated, the solution level is kept low enough that enough roots are above the solution so they get adequate oxygen. A hole is cut in the lid of the reservoir for each plant. There can be one to many plants per reservoir. Reservoir size can be increased as plant– size increases. A home made system can be constructed from plastic food containers or glass canning jars with aeration provided by an aquarium pump, aquarium airline tubing and aquarium valves. Clear containers are covered with aluminium foil, butcher paper, black plastic, or other material to exclude light, thus helping to eliminate the formation of algae. The nutrient solution is changed either on a schedule, such as once per week, or when the concentration drops below a certain level as determined with an electrical conductivity meter. Whenever the solution is depleted below a certain level, either water or fresh nutrient solution is added. A Mariotte's bottle, or a float valve, can be used to automatically maintain the solution level. In raft solution culture, plants are placed in a sheet of buoyant plastic that is floated on the surface of the nutrient solution. That way, the solution level never drops below the roots.

Continuous-flow Solution Culture

The nutrient film technique being used to grow various salad greens

In continuous-flow solution culture, the nutrient solution constantly flows past the roots. It is much easier to automate than the static solution culture because sampling and adjustments to the temperature and nutrient concentrations can be made in a large storage tank that has potential to serve thousands of plants. A popular variation is the nutrient film technique or NFT, whereby a very shallow stream of water containing all the dissolved nutrients required for plant growth is re-circulated past the bare roots of plants in a watertight thick root mat, which develops in the bottom of the channel and has an upper surface that, although moist, is in the air. Subsequent to this, an abundant supply of oxygen is provided to the roots of the plants. A properly designed NFT system is based on using the right channel slope, the right flow rate, and the right channel length. The main advantage of the NFT system over other forms of hydroponics is that the plant roots are exposed to adequate supplies of water, oxygen, and nutrients. In all other forms of production, there is a conflict between the supply of these requirements, since excessive or deficient amounts of one results in an imbalance of one or both of the others. NFT, because of its design, provides a system where all three requirements for healthy plant growth can be met at the same time, provided that the simple concept of NFT is always remembered and practised. The result of these advantages is that higher yields of high-quality produce are obtained over an extended period of cropping. A downside of NFT is that it has very little buffering against interruptions in the flow (e.g. power outages). But, overall, it is probably one of the more productive techniques.

The same design characteristics apply to all conventional NFT systems. While slopes along channels of 1:100 have been recommended, in practice it is difficult to build a base for channels that is sufficiently true to enable nutrient films to flow without ponding in locally depressed areas. As a consequence, it is recommended that slopes of 1:30 to 1:40 are used. This allows for minor irregularities in the surface, but, even with these slopes, ponding and water logging may occur. The slope may be provided by the floor, or benches or racks may hold the channels and provide the required slope. Both methods are used and depend on local requirements, often determined by the site and crop requirements.

As a general guide, flow rates for each gully should be 1 liter per minute. At planting, rates may be half this and the upper limit of 2 L/min appears about the maximum. Flow rates beyond these extremes are often associated with nutritional problems. Depressed growth rates of many crops have been observed when channels exceed 12 metres in length. On rapidly growing crops, tests have indicated that, while oxygen levels remain adequate, nitrogen may be depleted over the length of the gully. As a consequence, channel length should not exceed 10–15 metres. In situations where this is not possible, the reductions in growth can be eliminated by placing another nutrient feed halfway along the gully and halving the flow rates through each outlet.

Aeroponics

Aeroponics is a system wherein roots are continuously or discontinuously kept in an environment saturated with fine drops (a mist or aerosol) of nutrient solution. The method requires no substrate and entails growing plants with their roots suspended in a deep air or growth chamber with the roots periodically wetted with a fine mist of atomized nutrients. Excellent aeration is the main advantage of aeroponics.

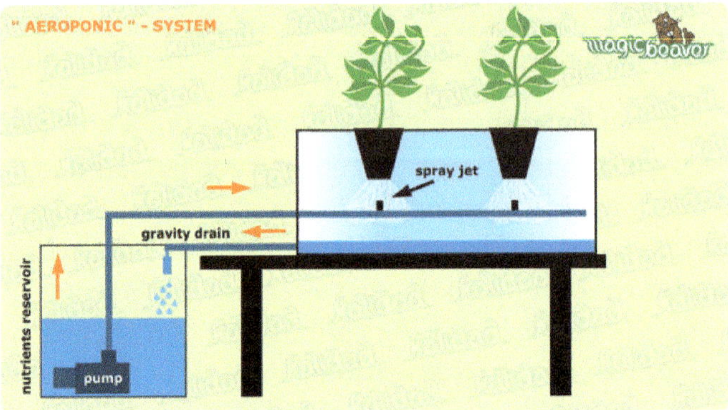

A diagram of the aeroponic technique.

Aeroponic techniques have proven to be commercially successful for propagation, seed germination, seed potato production, tomato production, leaf crops, and micro-greens. Since inventor Richard Stoner commercialized aeroponic technology in 1983, aeroponics has been implemented as an alternative to water intensive hydroponic systems worldwide. The limitation of hydroponics is the fact that 1 kilogram (2.2 lb) of water can only hold 8 milligrams (0.12 gr) of air, no matter whether aerators are utilized or not.

Another distinct advantage of aeroponics over hydroponics is that any species of plants can be grown in a true aeroponic system because the micro environment of an aeroponic can be finely controlled. The limitation of hydroponics is that only certain species of plants can survive for so long in water before they become waterlogged. The advantage of aeroponics is that suspended aeroponic plants receive 100% of the available oxygen and carbon dioxide to the roots zone, stems, and leaves, thus accelerating biomass growth and reducing rooting times. NASA research has shown that aeroponically grown plants have an 80% increase in dry weight biomass (essential minerals) compared to hydroponically grown plants. Aeroponics used 65% less water than hydroponics. NASA also concluded that aeroponically grown plants requires ¼ the nutrient input compared to hydroponics. Unlike hydroponically grown plants, aeroponically grown plants will not suffer transplant shock when transplanted to soil, and offers growers the ability to reduce the spread of disease and pathogens. Aeroponics is also widely used in laboratory studies of plant physiology and plant pathology. Aeroponic techniques have been given special attention from NASA since a mist is easier to handle than a liquid in a zero-gravity environment.

Fogponics

Fogponics is a derivation of aeroponics wherein the nutrient solution is aerosolized by a diaphragm vibrating at ultrasonic frequencies. Solution droplets produced by this method tend to be 5-10 μm in diameter, smaller than those produced by forcing a nutrient solution through pressurized nozzles, as in aeroponics. The smaller size of the droplets allows them to diffuse through the air more easily, and deliver nutrients to the roots without limiting their access to oxygen.

Passive Sub-irrigation

Passive sub-irrigation, also known as passive hydroponics or semi-hydroponics, is a method wherein plants are grown in an inert porous medium that transports water and fertilizer to the roots by capillary action from a separate reservoir as necessary, reducing labor and providing a constant supply of water to the roots. In the simplest method, the pot sits in a shallow solution of fertilizer and water or on a capillary mat saturated with nutrient solution. The various hydroponic media available, such as expanded clay and coconut husk, contain more air space than more traditional potting mixes, delivering increased oxygen to the roots, which is important in epiphytic plants such as orchids and bromeliads, whose roots are exposed to the air in nature. Additional advantages of passive hydroponics are the reduction of root rot and the additional ambient humidity provided through evaporations.

Ebb and Flow or Flood and Drain Sub-irrigation

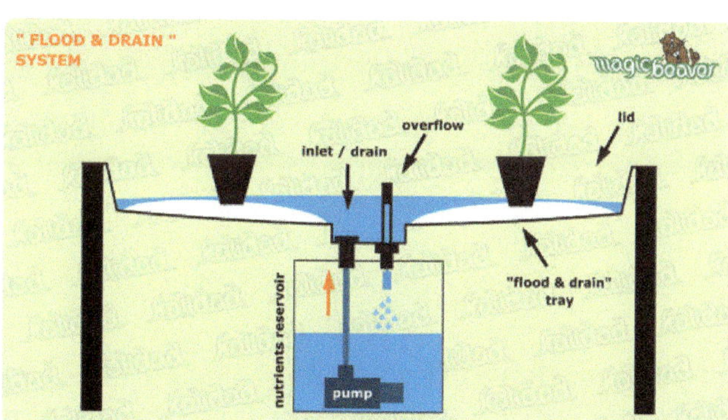

A Ebb and flow or flood and drain hydroponics system.

In its simplest form, there is a tray above a reservoir of nutrient solution. Either the tray is filled with growing medium (clay granules being the most common) and planted directly or pots of medium stand in the tray. At regular intervals, a simple timer causes a pump to fill the upper tray with nutrient solution, after which the solution drains back down into the reservoir. This keeps the medium regularly flushed with nutrients and air. Once the upper tray fills past the drain stop, it begins recirculating the water until the timer turns the pump off, and the water in the upper tray drains back into the reservoirs.

Run to Waste

In a run-to-waste system, nutrient and water solution is periodically applied to the medium surface. The method was invented in Bengal in 1946, for this reason it is sometimes referred to as "The Bengal System".

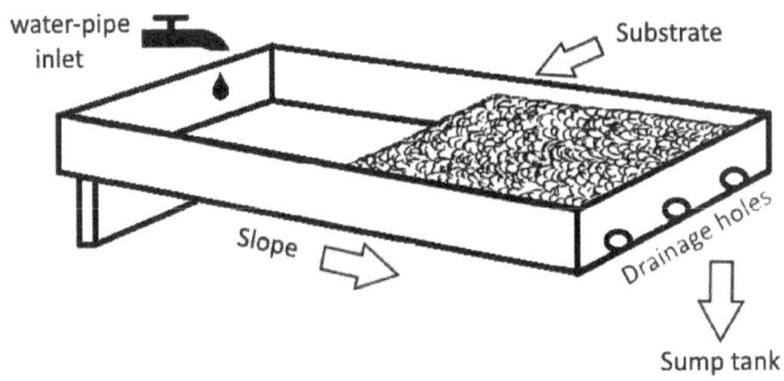

A run-to-waste hydroponics system referred to as "The Bengal System" after the region in northeastern India where it was invented (circa 1946–1948).

This method can be setup in various configurations. In its simplest form, a nutrient-and-water solution is manually applied one or more times per day to a container of inert growing media, such as rockwool, perlite, vermiculite, coco fibre, or sand. In a slightly more complex system, it is automated with a delivery pump, a timer and irrigation tubing to deliver nutrient solution with a delivery frequency that is governed by the key parameters of plant size, plant growing stage, climate, substrate, and substrate conductivity, pH, and water content.

In a commercial setting, watering frequency is multi-factorial and governed by computers or PLCs.

Commercial hydroponics production of large plants like tomatoes, cucumber, and peppers use one form or another of run-to-waste hydroponics.

In environmentally responsible uses, the nutrient rich waste is collected and processed through an on site filtration system to be used many times, making the system very productive.

The majority of bonsai are now grown in soil-free substrates (typically consisting of akadama, grit, diatomaceous earth and other inorganic components) and have their water and nutrients provided in a run-to-waste form.

Deep Water Culture

The hydroponic method of plant production by means of suspending the plant roots in a solution of nutrient-rich, oxygenated water. Traditional methods favor the use of plastic buckets and large containers with the plant contained in a net pot suspended from the centre of the lid and the roots suspended in the nutrient solution. The solution is oxygen saturated by an air pump combined with porous stones. With this method, the plants grow much faster because of the high amount of oxygen that the roots receive.

The Deep water culture technique being used to grow Hungarian wax peppers.

Top-fed Deep Water Culture

Top-fed deep water culture is a technique involving delivering highly oxygenated nutrient solution direct to the root zone of plants. While deep water culture involves the plant roots hanging down into a reservoir of nutrient solution, in top-fed deep water culture the solution is pumped from the reservoir up to the roots (top feeding). The water is released over the plant's roots and then runs back into the reservoir below in a constantly recirculating system. As with deep water culture, there is an airstone in the reservoir that pumps air into the water via a hose from outside the reservoir. The airstone helps add oxygen to the water. Both the airstone and the water pump run 24 hours a day.

The biggest advantage of top-fed deep water culture over standard deep water culture is increased growth during the first few weeks. With deep water culture, there is a time when the roots have not reached the water yet. With top-fed deep water culture, the roots get easy access to water from the beginning and will grow to the reservoir below much more quickly than with a deep water culture system. Once the roots have reached the reservoir below, there is not a huge advantage with top-fed deep water culture over standard deep water culture. However, due to the quicker growth in the beginning, grow time can be reduced by a few weeks.

Rotary

A rotary hydroponic garden is a style of commercial hydroponics created within a circular frame which rotates continuously during the entire growth cycle of whatever plant is being grown.

A Rotary hydroponic cultivation demonstration at the Belgian Pavilion Expo in 2015.

While system specifics vary, systems typically rotate once per hour, giving a plant 24 full turns within the circle each 24-hour period. Within the center of each rotary hydroponic garden is a high intensity grow light, designed to simulate sunlight, often with the assistance of a mechanized timer.

Each day, as the plants rotate, they are periodically watered with a hydroponic growth solution to provide all nutrients necessary for robust growth. Due to the plants continuous fight against gravity, plants typically mature much more quickly than when grown in soil or other traditional hydroponic growing systems. Due to the small foot print a rotary hydroponic system has, it allows for more plant material to be grown per sq foot of floor space than other traditional hydroponic systems.

Substrates

One of the most obvious decisions hydroponic farmers have to make is which medium they should use. Different media are appropriate for different growing techniques.

Expanded Clay Aggregate

Expanded clay pebbles.

Baked clay pellets are suitable for hydroponic systems in which all nutrients are carefully controlled in water solution. The clay pellets are inert, pH neutral and do not contain any nutrient value.

The clay is formed into round pellets and fired in rotary kilns at 1,200 °C (2,190 °F). This causes the clay to expand, like popcorn, and become porous. It is light in weight, and does not compact over time. The shape of an individual pellet can be irregular or uniform depending on brand and manufacturing process. The manufacturers consider expanded clay to be an ecologically sustainable and re-usable growing medium because of its ability to be cleaned and sterilized, typically by washing in solutions of white vinegar, chlorine bleach, or hydrogen peroxide (H2O2), and rinsing completely.

Another view is that clay pebbles are best not re-used even when they are cleaned, due to root growth that may enter the medium. Breaking open a clay pebble after a crop has been grown will reveal this growth.

Growstones

Growstones, made from glass waste, have both more air and water retention space than perlite and peat. This aggregate holds more water than parboiled rice hulls. Growstones by volume consist of 0.5 to 5% calcium carbonate – for a standard 5.1 kg bag of Growstones that corresponds to 25.8 to 258 grams of calcium carbonate. The remainder is soda-lime glass.

Coir Peat

Coco peat, also known as coir or coco, is the leftover material after the fibres have been removed from the outermost shell (bolster) of the coconut. Coir is a 100% natural grow and flowering medium. Coconut coir is colonized with trichoderma fungi, which protects roots and stimulates root growth. It is extremely difficult to over-water coir due to its perfect air-to-water ratio; plant roots thrive in this environment. Coir has a high cation exchange, meaning it can store unused minerals to be released to the plant as and when it requires it. Coir is available in many forms; most common is coco peat, which has the appearance and texture of soil but contains no mineral content.

Rice Husks

Rice husks, a hydroponic growing substrate option.

Parboiled rice husks (PBH) are an agricultural byproduct that would otherwise have little use. They decay over time, and allow drainage, and even retain less water than growstones. A study showed that rice husks did not affect the effects of plant growth regulators.

Perlite

Perlite, a hydroponic growing substrate option.

Perlite is a volcanic rock that has been superheated into very lightweight expanded glass pebbles. It is used loose or in plastic sleeves immersed in the water. It is also used in potting soil mixes to decrease soil density. Perlite has similar properties and uses to vermiculite but, in general, holds more air and less water. If not contained, it can float if flood and drain feeding is used. It is a fusion of granite, obsidian, pumice and basalt. This volcanic rock is naturally fused at high temperatures undergoing what is called "Fusionic Metamorphosis".

Vermiculite

Vermiculite close-up.

Like perlite, vermiculite is a mineral that has been superheated until it has expanded into light pebbles. Vermiculite holds more water than perlite and has a natural "wicking" property that can draw water and nutrients in a passive hydroponic system. If too much water and not enough air surrounds the plants roots, it is possible to gradually lower the medium's water-retention capability by mixing in increasing quantities of perlite.

Pumice

A pumice stone.

Like perlite, pumice is a lightweight, mined volcanic rock that finds application in hydroponics.

Sand

Sand is cheap and easily available. However, it is heavy, does not hold water very well, and it must be sterilized between uses.

Gravel

The same type that is used in aquariums, though any small gravel can be used, provided it is washed first. Indeed, plants growing in a typical traditional gravel filter bed, with water circulated using electric powerhead pumps, are in effect being grown using gravel hydroponics. Gravel is inexpensive, easy to keep clean, drains well and will not become waterlogged. However, it is also heavy, and, if the system does not provide continuous water, the plant roots may dry out.

Wood Fibre

Wood fibre, produced from steam friction of wood, is a very efficient organic substrate for hydroponics. It has the advantage that it keeps its structure for a very long time. Wood wool (i.e. wood slivers) have been used since the earliest days of the hydroponics research. However, more recent research suggests that wood fibre may have detrimental effects on "plant growth regulators".

Sheep Wool

Wool from shearing sheep is a little-used yet promising renewable growing medium. In a study comparing wool with peat slabs, coconut fibre slabs, perlite and rockwool slabs to grow cucumber plants, sheep wool had a greater air capacity of 70%, which decreased with use to a comparable 43%, and water capacity that increased from 23% to 44% with use. Using sheep wool resulted in the greatest yield out of the tested substrates, while application of a biostimulator consisting of humic acid, lactic acid and Bacillus subtilis improved yields in all substrates.

Rock Wool

Rock wool close-up.

Rock wool (mineral wool) is the most widely used medium in hydroponics. Rock wool is an inert substrate suitable for both run-to-waste and recirculating systems. Rock wool is made from molten rock, basalt or 'slag' that is spun into bundles of single filament fibres, and bonded into a medium capable of capillary action, and is, in effect, protected from most common microbiological degradation. Rock wool has many advantages and some disadvantages. The latter being the possible skin irritancy (mechanical) whilst handling (1:1000). Flushing with cold water usually brings relief. Advantages include its proven efficiency and effectiveness as a commercial hydroponic substrate. Most of the rock wool sold to date is a non-hazardous, non-carcinogenic material, falling under Note Q of the European Union Classification Packaging and Labeling Regulation (CLP).

Brick Shards

Brick shards have similar properties to gravel. They have the added disadvantages of possibly altering the pH and requiring extra cleaning before reuse.

Polystyrene Packing Peanuts

Polystyrene foam peanuts

Polystyrene packing peanuts are inexpensive, readily available, and have excellent drainage. However, they can be too lightweight for some uses. They are used mainly in closed-tube systems. Note that polystyrene peanuts must be used; biodegradable packing peanuts will decompose into a sludge. Plants may absorb styrene and pass it to their consumers; this is a possible health risk.

Nutrient Solutions

Inorganic Hydroponic Solutions

The formulation of hydroponic solutions is an application of plant nutrition, with nutrient deficiency symptoms mirroring those found in traditional soil based agriculture. However, the underlying chemistry of hydroponic solutions can differ from soil chemistry in many significant ways. Important differences include:

- Unlike soil, hydroponic nutrient solutions do not have cation-exchange capacity (CEC) from clay particles or organic matter. The absence of CEC means the pH and nutrient concentrations can change much more rapidly in hydroponic setups than is possible in soil.
- Selective absorption of nutrients by plants often imbalances the amount of counterions in solution. This imbalance can rapidly affect solution pH and the ability of plants to absorb nutrients of similar ionic charge. For instance, nitrate anions are often consumed rapidly by plants to form proteins, leaving an excess of cations in solution. This cation im-

balance can lead to deficiency symptoms in other cation based nutrients (e.g. Mg^{2+}) even when an ideal quantity of those nutrients are dissolved in the solution.

- Depending the on pH, and/or the presence of water contaminants, nutrients, such as iron, can precipitate from the solution and become unavailable to plants. Routine adjustments to pH, buffering the solution, and/or the use of chelating agents is often necessary.

As in conventional agriculture, nutrients should be adjusted to satisfy Liebig's law of the minimum for each specific plant variety. Nevertheless, generally acceptable concentrations for nutrient solutions exist, with minimum and maximum concentration ranges for most plants being somewhat similar. Most nutrient solutions are mixed to have concentrations between 1,000 and 2,500 ppm. Acceptable concentrations for the individual nutrient ions, which comprise that total ppm figure, are summarized in the following table. For essential nutrients, concentrations below these ranges often lead to nutrient deficiencies while exceeding these ranges can lead to nutrient toxicity. Optimum nutrition concentrations for plant varieties are found empirically by experience and/or by plant tissue tests.

Element	Role	Ionic form(s)	Low range (ppm)	High range (ppm)	Common Sources	Comment
Nitrogen	Essential macronutrient	NO_3^- and/ or NH_4^+	100	1000	KNO_3, NH_4NO_3, $Ca(NO_3)_2$, HNO_3, $(NH_4)_2SO_4$, and $(NH_4)_2HPO_4$	NH_4^+ interferes with Ca^{2+} uptake and can be toxic to plants if used as a major nitrogen source. A 3:1 ratio of NO_3^- to NH_4^+ is sometimes recommended to balance pH during nitrogen absorption.
Potassium	Essential macronutrient	K^+	100	400	KNO_3, K_2SO_4, KCl, KOH, K_2CO_3, K_2HPO_4, and K_2SiO_3	High concentrations interfere with the function Fe, Mn, and Zn. Zinc deficiencies often are the most apparent.
Phosphorus	Essential macronutrient	PO_4^{3-}	30	100	K_2HPO_4, KH_2PO_4, $NH_4H_2PO_4$, H_3PO_4, and $Ca(H_2PO_4)_2$	Excess NO_3^- tends to inhibit PO_4^{3-} absorption. The ratio of iron to PO_4^{3-} can affect co-preciptiation reactions.
Calcium	Essential macronutrient	Ca^{2+}	200	500	$Ca(NO_3)_2$, $Ca(H_2PO_4)$, $CaSO_4$, $CaCl_2$	Excess Ca^{2+} inhibits Mg^{2+} uptake.
Magnesium	Essential macronutrient	Mg^{2+}	50	100	$MgSO_4$ and $MgCl_2$	Should not exceed Ca^{2+} concentration due to competitive uptake.
Sulfur	Essential macronutrient	SO_4^{2-}	50	1000	$MgSO_4$, K_2SO_4, $CaSO_4$, H_2SO_4, $(NH_4)_2SO_4$, $ZnSO_4$, $CuSO_4$, $FeSO_4$, and $MnSO_4$	Unlike most nutrients, plants can tolerate a high concentration of the SO_4^{2-}, selectively absorbing the nutrient as needed. Undesirable counterion affects still apply however.

Iron	Essential micronutrient	Fe^{3+} and Fe^{2+}	2	5	FeDTPA, FeEDTA, iron citrate, iron tartrate, $FeCl_3$, and $FeSO_4$	pH values above 6.5 greatly decreases iron solubility. Chelating agents (e.g. DTPA, citric acid, or EDTA) are often added to increase iron solubility over a greater pH range.
Zinc	Essential micronutrient	Zn^{2+}	0.05	1	$ZnSO_4$	Excess zinc is highly toxic to plants but is essential for plants at low concentrations.
Copper	Essential micronutrient	Cu^{2+}	0.01	1	$CuSO_4$	Plant sensitivity to copper is highly variable. 0.1 ppm can cause toxic to some plants while a concentration up to 0.5 ppm for many plants is often considered ideal.
Manganese	Essential micronutrient	Mn^{2+}	0.5	1	$MnSO_4$ and $MnCl_2$	Uptake is enhanced by high PO_4^{3-} concentrations.
Boron	Essential micronutrient	$B(OH)_4^-$	0.3	10	H_3BO_3, and $Na_2B_4O_7$	An essential nutrient, however, some plants are highly sensitive to boron (e.g. toxic affects are apparent in citrus trees at 0.5 ppm).
Molybdenum	Essential micronutrient	MoO_4^-	0.001	0.05	$(NH_4)_6Mo_7O_{24}$ and Na_2MoO_4	A component of the enzyme nitrate reductase and required by rhizobia for nitrogen fixation.
Nickel	Essential micronutrient	Ni^{2+}	0.057	1.5	$NiSO_4$ and $NiCO_3$	Essential to many plants (e.g. legumes and some grain crops). Also used in the enzyme urease.
Chlorine	Variable micronutrient	Cl^-	0	Highly variable	KCl, $CaCl_2$, $MgCl_2$, and NaCl	Can interfere with NO_3^- uptake in some plants but can be beneficial in some plants (e.g. in asparagus at 5 ppm). Absent in conifers, ferns, and most bryophytes.
Aluminum	Variable micronutrient	Al^{3+}	0	10	$Al_2(SO_4)_3$	Essential for some plants (e.g. peas, maize, sunflowers, and cereals). Can be toxic to some plants below 10 ppm. Sometimes used to produce flower pigments (e.g. by Hydrangeas).
Silicon	Variable micronutrient	SiO_3^{2-}	0	140	K_2SiO_3, Na_2SiO_3, and H_2SiO_3	Present in most plants, abundant in cereal crops, grasses, and tree bark. Evidence that SiO_3^{2-} improves plant disease resistance exists.

Titanium	Variable micronutrient	Ti^{3+}	0	5	H_4TiO_4	Might be essential but trace Ti^{3+} is so ubiquitous that its addition is rarely warranted. At 5 ppm favorable growth effects in some crops are notable (e.g. pineapple and peas).
Cobalt	Non-essential micronutrient	Co^{2+}	0	0.1	$CoSO_4$	Required by rhizobia, important for legume root nodulation.
Sodium	Non-essential micronutrient	Na^+	0	Highly variable	Na_2SiO_3, Na_2SO_4, NaCl, $NaHCO_3$, and NaOH	Na^+ can partially replace K^+ in some plant functions but K^+ is still an essential nutrient.
Vanadium	Non-essential micronutrient	VO^{2+}	0	Trace,	$VOSO_4$	Beneficial for rhizobial N_2 fixation.
Lithium	Non-essential micronutrient	Li^+	0		Li_2SO_4, LiCl, and LiOH	Li^+ can increase the chlorophyll content of some plants (e.g. potato and pepper plants).

Organic Hydroponic Solutions

Organic fertilizers can be used to supplement or entirely replace the inorganic compounds used in conventional hydroponic solutions. However, using organic fertilizers introduces a number of challenges that are not easily resolved. Examples include:

- organic fertilizers are highly variable in their nutritional compositions. Even similar materials can differ significantly based on their source (e.g. the quality of manure varies based on an animal's diet).
- organic fertilizers are often sourced from animal byproducts, making disease transmission a serious concern for plants grown for human consumption or animal forage.
- organic fertilizers are often particulate and can clog substrates or other growing equipment. Sieving and/or milling the organic materials to fine dusts is often necessary.
- some organic materials (i.e. particularly manures and offal) can further degrade to emit foul odors.

Nevertheless, if precautions are taken, organic fertilizers can be used successfully in hydroponics.

Organically Sourced Macronutrients

Examples of suitable materials, with their average nutritional contents tabulated in terms of percent dried mass, are listed in the following table.

Organic material	N	P_2O_5	K_2O	CaO	MgO	SO_2	**Comment**
Bloodmeal	13.0%	2.0%	1.0%	0.5%	–	–	
Bone ashes	–	35.0%	–	46.0%	1.0%	0.5%	

Bonemeal	4.0%	22.5%	–	33.0%	0.5%	0.5%	
Hoof / Horn meal	14.0%	1.0%	–	2.5%	–	2.0%	
Fishmeal	9.5%	7.0%	–	0.5%	–	–	
Wool waste	3.5%	0.5%	2.0%	0.5%	–	–	
Wood ashes	–	2.0%	5.0%	33.0%	3.5%	1.0%	
Cottonseed ashes	–	5.5%	27.0%	9.5%	5.0%	2.5%	
Cottonseed meal	7.0%	3.0%	2.0%	0.5%	0.5%	–	
Dried locust or grasshopper	10.0%	1.5%	0.5%	0.5%	–	–	
Leather waste	5.5% to 22%	–	–	–	–	–	Milled to a fine dust.
Kelp meal, liquid seaweed	1%	–	12%	–	–	–	Commercial products available.
Poultry manure	2% to 5%	2.5% to 3%	1.3% to 3%	4.0%	1.0%	2.0%	A liquid compost which is sieved to remove solids and checked for pathogens.
Sheep manure	2.0%	1.5%	3.0%	4.0%	2.0%	1.5%	Same as poultry manure.
Goat manure	1.5%	1.5%	3.0%	2.0%	–	–	Same as poultry manure.
Horse manure	3% to 6%	1.5%	2% to 5%	1.5%	1.0%	0.5%	Same as poultry manure.
Cow manure	2.0%	1.5%	2.0%	4.0%	1.1%	0.5%	Same as poultry manure.
Bat guano	8.0%	40%	29%	Trace	Trace	Trace	High in micronutrients. Commercially available.
Bird guano	13%	8%	20%	Trace	Trace	Trace	High in micronutrients. Commercially available.

Organically Sourced Micronutrients

Micronutrients can be sourced from organic fertilizers as well. For example, composted pine bark is high in manganese and is sometimes used to fulfill that mineral requirement in hydroponic solutions. To satisfy requirements for National Organic Programs, pulverized, unrefined minerals (e.g. Gypsum, Calcite, and glauconite) can also be added to satisfy a plant's nutritional needs.

Additives

In addition to chelating agents, humic acids can be added to increase nutrient uptake.

Tools

Common Equipment

Managing nutrient concentrations and pH values within acceptable ranges is essential for successful hydroponic horticulture. Common tools used to manage hydroponic solutions include:

- Electrical conductivity meters, a tool which estimates nutrient ppm by measuring how well a solution transmits an electric current.

- pH meter, a tool that uses an electric current to determine the concentration of hydrogen ions in solution.
- Litmus paper, disposable pH indicator strips that determine hydrogen ion concentrations by color changing chemical reaction.
- Graduated cylinders or measuring spoons to measure out premixed, commercial hydroponic solutions.

Advanced Equipment

Advanced equipment can also be used to perform accurate chemical analyses of nutrient solutions. Examples include:

- Balances for accurately measuring materials.
- Laboratory glassware, such as burettes and pipettes, for performing titrations.
- Colorimeters for solution tests which apply the Beer–Lambert law.

Using advanced equipment for hydroponic solutions can be beneficial to growers of any background because nutrient solutions are often reusable. Because nutrient solutions are virtually never completely depleted, and should never be due to the unacceptably low osmotic pressure that would result, re-fortification of old solutions with new nutrients can save growers money and can control point source pollution, a common source for the eutrophication of nearby lakes and streams.

Software

Although pre-mixed concentrated nutrient solutions are generally purchased from commercial nutrient manufacturers by hydroponic hobbyists and small commercial growers, several tools exist to help anyone prepare their own solutions without extensive knowledge about chemistry. The free and open source tools HydroBuddy and HydroCal have been created by professional chemists to help any hydroponics grower prepare their own nutrient solutions. The first program is available for Windows, Mac and Linux while the second one can be used through a simple JavaScript interface. Both programs allow for basic nutrient solution preparation although HydroBuddy provides added functionality to use and save custom substances, save formulations and predict electrical conductivity values.

Mixing Solutions

Often mixing hydroponic solutions using individual salts is impractical for hobbyists and/or small-scale commercial growers because commercial products are available at reasonable prices. However, even when buying commercial products, multi-component fertilizers are popular. Often these products are bought as three part formulas which emphasize certain nutritional roles. For example, solutions for vegetative growth (i.e. high in nitrogen), flowering (i.e. high in potassium and phosphorus), and micronutrient solutions (i.e. with trace minerals) are popular. The timing and application of these multi-part fertilizers should coincide with a plant's growth stage. For example, at the end of an annual plant's life cycle, a plant should be restricted from high nitrogen fertilizers. In most plants, nitrogen restriction inhibits vegetative growth and helps induce flowering.

Advancements

With pest problems reduced and nutrients constantly fed to the roots, productivity in hydroponics is high; however, growers can further increase yield by manipulating a plant's environment by constructing sophisticated growrooms.

CO_2 Enrichment

To increase yield further, some sealed greenhouses inject CO_2 into their environment to help improve growth and plant fertility.

References

- Douglas, James Sholto (1975). Hydroponics: The Bengal System (5th ed.). New Dehli: Oxford University Press. ISBN 9780195605662.
- A Dictionary of Chemistry (6 ed.), edited by John Daintith, Publisher: Oxford University Press, 2008, ISBN 9780199204632: "greenhouse effect"
- Brian Shmaefsky (2004). Favorite demonstrations for college science: an NSTA Press journals collection. NSTA Press. p. 57. ISBN 978-0-87355-242-4.
- Griffith, Anna N. (1985), Collins Guide to Alpines and Rock Garden Plants, London: Collins, pp. 20–21, ISBN 978-0-907486-81-7
- Chittaranjan Kole (29 December 2010). Wild Crop Relatives - Genomic and Breeding Resources: Temperate Fruits. Springer. p. 59. ISBN 978-3-642-16056-1. Retrieved 25 March 2011.
- Thorpe, David, 13/11/2014, The One Planet Life: A Blueprint For Low Impact Development, Taylor and Francis, p376, ISBN 9781317625902
- Douglas John McConnell (2003). The Forest Farms of Kandy: And Other Gardens of Complete Design. p. 1. ISBN 9780754609582.
- Ryrie, Charlie (2004). The Cottage Garden: How to Plan and Plant a Garden That Grows Itself. Collins & Brown. p. 7. ISBN 1-84340-216-5.
- Scott-James, Anne; Osbert Lancaster (2004). The Pleasure Garden: An Illustrated History of British Gardening. Frances Lincoln Publishers. p. 80. ISBN 978-0-7112-2360-8.
- A Biographical Dictionary of British Architects, 1600–1840, Howard Colvin, Yale University Press, 2008 ISBN 0-300-12508-9, p 659
- Lloyd, Christopher; Richard Bird (1999). The Cottage Garden. Jacqui Hurst. Dorling Kindersley. pp. 6–9. ISBN 978-0-7513-0702-3.
- Boults, Elizabeth and Chip Sullivan (2010). Illustrated History of Landscape Design. John Wiley and Sons. p. 175. ISBN 0-470-28933-3.
- Hemenway, Toby (2009). Gaia's Garden: A Guide to Home-Scale Permaculture. Chelsea Green Publishing. pp. 84-85. ISBN 978-1-60358-029-8.
- Gericke, William F. (1940). The Complete Guide to Soilless Gardening (1st ed.). London: Putnam. pp. 9–10, 38 & 84. ISBN 9781163140499.

- Douglas, James Sholto (1975). Hydroponics: The Bengal System (5th ed.). New Dehli: Oxford University Press. p. 10. ISBN 9780195605662.
- Sholto Douglas, James (1985). Advanced guide to hydroponics: (soiless cultivation). London: Pelham Books. pp. 169–187, 289–320, & 345–351. ISBN 9780720715712.
- J. Benton, Jones (2004). Hydroponics: A Practical Guide for the Soilless Grower (2nd ed.). Newyork: Taylor & Francis. pp. 29–70 & 225–229. ISBN 9780849331671.

4

Plant Genetics: An Integrated Study

Plant genetics is a branch of genetic science that is involved in the mapping and sequencing of the plant genome as well as producing genetically altered plants (GMOs). The chapter delves deep into the predominant methods of modifying plants through the use of the gene gun and the agrobacterium method. A section of the content elaborates on genetically modified trees and crops.

Plant Genetics

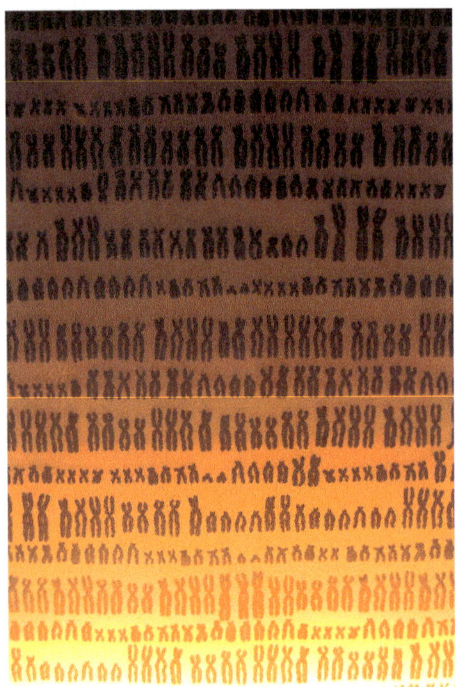

An image of multiple chromosomes, making up a genome

Plant genetics is a very broad term. There are many facets of genetics in general, and of course there are many facets to plants. The definition of genetics is the branch of biology that deals with heredity, especially the mechanisms of hereditary transmission and the variation of inherited characteristics among similar or related organisms. And the definition of a plant is any of various photosynthetic, eukaryotic, multicellular organisms of the kingdom Plantae characteristically producing embryos, containing chloroplasts, having cell walls which contain cellulose, and lacking the power of locomotion. Although there has been a revolution in the biological sciences in the past twenty years, there is still a great deal that remains to be discovered. The completion of the sequencing of the genomes of rice and some agriculturally and scientifically important plants (for example Physcomitrella patens) has increased the possibilities of plant genetic research immeasurably.

Features of Plant Biology

Plant genetics is different from that of animals in a few ways. Like mitochondria, chloroplasts have their own DNA, complicating pedigrees somewhat. Like animals, plants have somatic mutations regularly, but these mutations can contribute to the germ line with ease, since flowers develop at the ends of branches composed of somatic cells. People have known of this for centuries, and mutant branches are called "sports". If the fruit on the sport is economically desirable, a new cultivar may be obtained.

Some plant species are capable of self-fertilization, and some are nearly exclusively self-fertilizers. This means that a plant can be both mother and father to its offspring, a rare occurrence in animals. Scientists and hobbyists attempting to make crosses between different plants must take special measures to prevent the plants from self-fertilizing.

Plants are generally more capable of surviving, and indeed flourishing, as polyploids. Polyploidy, the presence of extra sets of chromosomes, is not usually compatible with life in animals. In plants, polyploid individuals are created frequently by a variety of processes, and once created usually cannot cross back to the parental type. Polyploid individuals, if capable of self-fertilizing, can give rise to a new genetically distinct lineage, which can be the start of a new species. This is often called "instant speciation". Polyploids generally have larger fruit, an economically desirable trait, and many human food crops, including wheat, maize, potatoes, peanuts, strawberries and tobacco, are either accidentally or deliberately created polyploids.

Hybrids between plant species are easy to create by hand-pollination, and may be more successful on average than hybrids between animal species. Often tens of thousands of offspring from a single cross are raised and tested to obtain a single individual with desired characteristics. People create hybrids for economic and aesthetic reasons, especially with orchids.

DNA

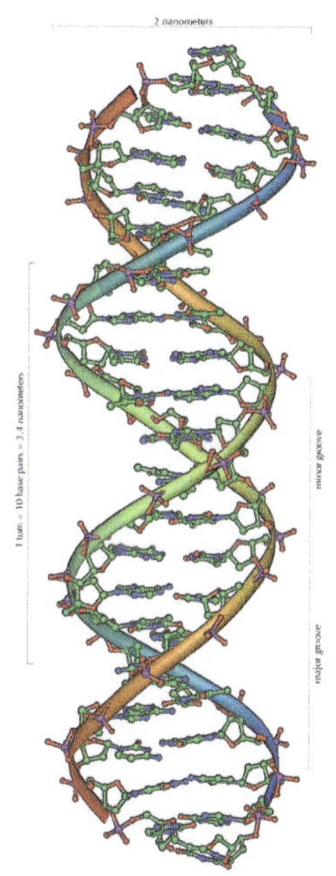

The structure of part of a DNA double helix

Deoxyribonucleic acid (DNA) is a nucleic acid that contains the genetic instructions used in the development and functioning of all known living organisms and some viruses. The main role of DNA molecules is the long-term storage of information. DNA is often compared to a set of blueprints or a recipe, or a code, since it contains the instructions needed to construct other components of cells, such as proteins and RNA molecules. The DNA segments that carry this genetic information are called genes, but other DNA sequences have structural purposes, or are involved in regulating the use of this genetic information. Geneticists, including plant geneticists, use this sequencing of DNA to their advantage as they splice and delete certain genes and regions of the DNA molecule to produce a different or desired genotype and thus, also producing a different phenotype.

Gregor Mendel

Gregor Mendel was an Augustinian priest and scientist born on 20 July 1822 in Austria-Hungary and is well known for discovering genetics. He went to the Abbey of St. Thomas in Brno. He is often called the father of genetics for his study of the inheritance of certain traits in pea plants. Mendel showed that the inheritance of these traits follows particular laws, which were later named after him. The significance of Mendel's work was not recognized until the turn of the 20th century. Its rediscovery prompted the foundation of the discipline of genetics allows geneticists today to accurately predict the outcome of such crosses and in determining the phenotypical effects of the crosses. He died on 6 January 1884 from chronic nephritis.

Modern Ways to Genetically Modify Plants

There are two predominant procedures of transforming genes in organisms: the "Gene gun" method and the *Agrobacterium* method.

"Gene Gun" Method

The "Gene Gun" method is also referred to as "biolistics" (ballistics using biological components). This technique is used for in vivo (within a living organism) transformation and has been especially useful in transforming monocot species like corn and rice. This approach literally shoots genes into plant cells and plant cell chloroplasts. DNA is coated onto small particles of gold or tungsten approximately two micrometres in diameter. The particles are placed in a vacuum chamber and the plant tissue to be engineered is placed below the chamber. The particles are propelled at high velocity using a short pulse of high pressure helium gas, and hit a fine mesh baffle placed above the tissue while the DNA coating continues into any target cell or tissue.

Agrobacterium Method

Transformation via *Agrobacterium* has been successfully practiced in dicots, i.e. broadleaf plants, such as soybeans and tomatoes, for many years. Recently it has been adapted and is now effective in monocots like grasses, including corn and rice. In general, the *Agrobacterium* method is considered preferable to the gene gun, because of a greater frequency of single-site insertions of the foreign DNA, which allows for easier monitoring. In this method, the tumor inducing (Ti) region is removed from the T-DNA (transfer DNA) and replaced with the desired gene and a marker, which is then inserted into the organism. This may involve direct inoculation of the tissue with a culture of transformed Agrobacterium, or inoculation following treatment with micro-projectile bombardment, which wounds the tissue. Wounding of the target tissue causes the release of phenolic compounds by the plant, which induces invasion of the tissue by Agrobacterium. Because of this, microprojectile bombardment often increases the efficiency of infection with Agrobacterium. The marker is used to find the organism which has successfully taken up the desired gene. Tissues of the organism are then transferred to a medium containing an antibiotic or herbicide, depending on which marker was used. The *Agrobacterium* present is also killed by the antibiotic. Only tissues expressing the marker will survive and possess the gene of interest. Thus, subsequent steps in the process will only use these surviving plants. In order to obtain whole plants from these tissues, they are grown under controlled environmental conditions in tissue culture. This is a process of a series of media, each containing nutrients and hormones. Once the plants are grown and produce seed, the process of evaluating the progeny begins. This process entails selection of the seeds with the desired traits and then retesting and growing to make sure that the entire process has been completed successfully with the desired results.

Genetically Engineered Crops

The use of genetically engineered crops has helped many farmers deal with pest problems that reduce their crop production. The impact of pest-resistant crops has led to a much higher yield for farmers in today's world. They can use less pesticides which reduces the chemicals that they put into the ground. Certain engineered crops have led to farmers all over the world and in the United States to increase crop yield exponentially in recent years. Farmers can use a glyphosate herbicide to kill weeds, yet the genetically engineered corn is resistant to the herbicide and is left unaffected. Thus, fields are produced that are virtually weed free. Genetically engineered crops can also benefit farmers when dealing with potentially harmful viruses and bacteria. In the 1990s a mutant strain of virus was decimating the commercial corn fields of the United States. Scientists found a virus resistant strain of maize in the highlands of Mexico and extracted the part of the maize's genome that coded for resistance against the virus and incorporated it into their existing strain of commercial corn. This allowed the commercial strain to produce progeny that were resistant to the virus. Thus, the crops were saved from decimation.

Potential Detrimental Effects of Genetically Engineered Plants

According to John E. Berringer the outcome of releasing genetically modified organisms into the environment is still not known.

Gene Gun

A gene gun or a biolistic particle delivery system, originally designed for plant transformation, is a device for injecting cells with genetic information; the inserted genetic material are termed transgenes. The payload is an elemental particle of a heavy metal coated with plasmid DNA. This technique is often simply referred to as bioballistics or biolistics.

PDS-1000/He Particle Delivery System

This device is able to transform almost any type of cell, including plants, and is not limited to genetic material of the nucleus: it can also transform organelles, including plastids.

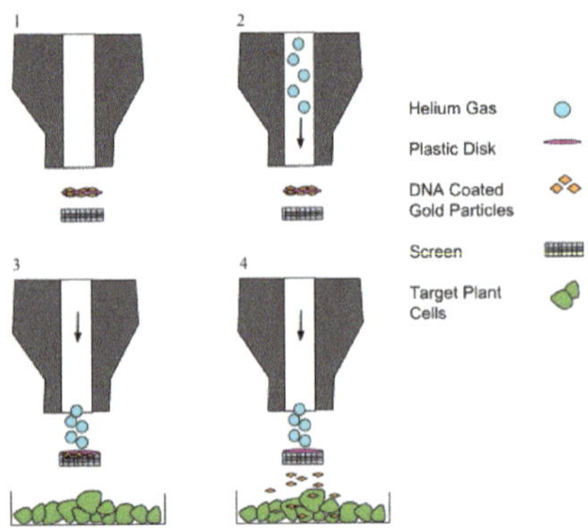

A gene gun is used for injecting cells with genetic information, it is also known as biolistic particle delivery system. Gene guns can be used effectively on most cells but are mainly used on plant cells. Step 1 The gene gun apparatus is ready to fire. Step 2 When the gun is turned on and the helium flows through. Step 3 The helium moving the disk with DNA coated particles toward the screen. Step 4 The helium having pushed the particles moving through the screen and moving to the target cells to transform the cells.

Gene Gun Design

The gene gun was originally a Crosman air pistol modified to fire dense tungsten particles. It was invented by John C Sanford, Ed Wolf and Nelson Allen at Cornell University, and Ted Klein of DuPont, between 1983 and 1986. The original target was onions (chosen for their large cell size) and it was used to deliver particles coated with a marker gene. Genetic transformation was then proven when the onion tissue expressed the gene.

The earliest custom manufactured gene guns (fabricated by Nelson Allen) used a 22 caliber nail gun cartridge to propel an extruded polyethylene cylinder (bullet) down a 22 cal. Douglas barrel.

A droplet of the tungsten powder and genetic material was placed on the bullet and shot down the barrel at a lexan "stopping" disk with a petri dish below. The bullet welded to the disk and the genetic information blasted into the sample in the dish with a doughnut effect (devastation in the middle, a ring of good transformation and little around the edge). The gun was connected to a vacuum pump and was under vacuum while firing. The early design was put into limited production by a Rumsey-Loomis (a local machine shop then at Mecklenburg Rd in Ithaca, NY, USA). Later the design was refined by removing the "surge tank" and changing to nonexplosive propellants. DuPont added a plastic extrusion to the exterior to visually improve the machine for mass production to the scientific community. Biorad contracted with Dupont to manufacture and distribute the device. Improvements include the use of helium propellant and a multi-disk-collision delivery mechanism. Other heavy metals such as gold and silver are also used. Gold may be favored because it has better uniformity than tungsten and tungsten can be toxic to cells, but its use may be limited due to availability and cost.

Biolistic Construct Design

A construct is a piece of DNA inserted into the target's genome, including parts that are intended to be removed later. All biolistic transformations require a construct to proceed and while there is great variation among biolistic constructs, they can be broadly sorted into two categories: those which are designed to transform eukaryotic nuclei, and those designed to transform prokaryotic-type genomes such as mitochondria, plasmids or plastids.

Those meant to transform prokaryotic genomes generally have the gene or genes of interest, at least one promoter and terminator sequence, and a reporter gene; which is a gene used to ease detection or removal of those cells which didn't integrate the construct into their DNA. These genes may each have their own promoter and terminator, or be grouped to produce multiple gene products from one transcript, in which case binding sites for translational machinery should be placed between each to ensure maximum translational efficiency. In any case the entire construct is flanked by regions called border sequences which are similar in sequence to locations within the genome, this allows the construct to target itself to a specific point in the existing genome.

Constructs meant for integration into a eukaryotic nucleus follow a similar pattern except that: the construct contains no border sequences because the sequence rearrangement that prokaryotic constructs rely on rarely occurs in eukaryotes; and each gene contained within the construct must be expressed by its own copy of a promoter and terminator sequence.

Though the above designs are generally followed, there are exceptions. For example, the construct might include a Cre-Lox system to selectively remove inserted genes; or a prokaryotic construct may insert itself downstream of a promoter, allowing the inserted genes to be governed by a promoter already in place and eliminating the need for one to be included in the construct.

Application

Gene guns are so far mostly applied for plant cells. However, there is much potential use in humans and other animals as well.

Plants

The target of a gene gun is often a callus of undifferentiated plant cells growing on gel medium in a Petri dish. After the gold particles have impacted the dish, the gel and callus are largely disrupted. However, some cells were not obliterated in the impact, and have successfully enveloped a DNA coated gold particle, whose DNA eventually migrates to and integrates into a plant chromosome.

Cells from the entire Petri dish can be re-collected and selected for successful integration and expression of new DNA using modern biochemical techniques, such as a using a tandem selectable gene and northern blots.

Selected single cells from the callus can be treated with a series of plant hormones, such as auxins and gibberellins, and each may divide and differentiate into the organized, specialized, tissue cells of an entire plant. This capability of total re-generation is called totipotency. The new plant that originated from a successfully shot cell may have new genetic (heritable) traits.

The use of the gene gun may be contrasted with the use of *Agrobacterium tumefaciens* and its Ti plasmid to insert genetic information into plant cells. See transformation for different methods of transformation in different species.

Humans and Other Animals

Gene guns have also been used to deliver DNA vaccines.

The delivery of plasmids into rat neurons through the use of a gene gun, specifically DRG neurons, is also used as a pharmacological precursor in studying the effects of neurodegenerative diseases such as Alzheimer's disease.

The gene gun has become a common tool for labeling subsets of cells in cultured tissue. In addition to being able to transfect cells with DNA plasmids coding for fluorescent proteins, the gene gun can be adapted to deliver a wide variety of vital dyes to cells.

Gene gun bombardment has also been used to transform *Caenorhabditis elegans*, as an alternative to microinjection.

Advantages

Biolistics has proven to be a versatile method of genetic modification and it is generally preferred to engineer transformation-resistant crops, such as cereals. Notably, *Bt* maize is a product of biolistics. Plastid transformation has also seen great success with particle bombardment when compared to other current techniques, such as *Agrobacterium* mediated transformation, which have difficulty targeting the vector to and stably expressing in the chloroplast. In addition, there are no reports of a chloroplast silencing a transgene inserted with a gene gun. Additionally, with only one firing of a gene gun, a skilled technician can generate two transformed organisms. This technology has even allowed for modification of specific tissues *in situ*, although this is likely to damage large numbers of cells and transform only some, rather than all, cells of the tissue.

Limitations

However, biolistics introduces the construct randomly into the target cells. Thus the altered DNA sequences may be transformed into whatever genomes are present in the cell, be they nuclear, mitochondrial, plasmid or any others, in any combination, though proper construct design may mitigate this. Another issue is that the gene inserted may be overexpressed when the construct is inserted multiple times in either the same or different locations of the genome. This is due to the ability of the constructs to give and take genetic information from other constructs, causing some to carry no transgene and others to carry multiple copies; the number of copies inserted depends on both how many copies of the transgene an inserted construct has, and how many were inserted. Also, because eukaryotic constructs rely on illegitimate recombination, a process by which the transgene is integrated into the genome without similar genetic sequences, and not homologous recombination, which inserts at similar sequences, they cannot be targeted to specific locations within the genome.

Genetically Modified Tree

A genetically modified tree (GMt, GM tree, genetically engineered tree, GE tree or transgenic tree) is a tree whose DNA has been modified using genetic engineering techniques. In most cases the aim is to introduce a novel trait to the plant which does not occur naturally within the species. Examples include resistance to certain pests, diseases, environmental conditions, and herbicide tolerance, or the alteration of lignin levels in order to reduce pulping costs.

Technician checks on genetically modified peach and apple "orchards". Each dish holds experimental trees grown from lab-cultured cells to which researchers have given new genes. Source: USDA.

Genetically modified forest trees are not yet approved ("deregulated") for commercial use, with the exception of insect-resistant poplar trees in China. and one case of GM Eucalyptus in Brazil.

Several genetically modified forest tree species are undergoing field trials for deregulation, and much of the research is being carried out by the pulp and paper industry, primarily with the intention of increasing the productivity of existing tree stock. Certain genetically modified orchard tree species have been deregulated for commercial use in the United States including the papaya and plum. The development, testing and use of GM trees remains at an early stage in comparison to GM crops.

Research

Research into genetically modified trees has been ongoing since 1988. Concerns surrounding the biosafety implications of releasing genetically modified trees into the wild have held back regulatory approval of GM forest trees. This concern is exemplified in the Convention on Biological Diversity's stance:

The Conference of the Parties, Recognising the uncertainties related to the potential environmental and socio-economic impacts, including long term and trans-boundary impacts, of genetically modified trees on global forest biological diversity, as well as on the livelihoods of indigenous and local communities, and given the absence of reliable data and of capacity in some countries to undertake risk assessments and to evaluate those potential impacts, recommends parties to take a precautionary approach when addressing the issue of genetically modified trees.

A precondition for further commercialization of GM forest trees is likely to be their complete sterility. Plantation trees remain phenotypically similar to their wild cousins in that most are the product of no more than three generations of artificial selection, therefore, the risk of transgene escape by pollination with compatible wild species is high. One of the most credible science based concerns with GM trees is their potential for wide dispersal of seed and pollen. The fact that pine pollen travels long distances is well established, moving up to 3,000 kilometers from its source. Additionally, many tree species reproduce for a long time before being harvested. In combination these factors have led some to believe that GM trees are worthy of special environmental considerations over GM crops. Ensuring sterility for GM trees has proven elusive, but efforts are being made. While tree geneticist Steve Strauss predicted that complete containment might be possible by 2020, many questions remain.

Proposed Uses

GM trees under experimental development have been modified with traits intended to provide benefit to industry, foresters or consumers. Due to high regulatory and research costs, the majority of genetically modified trees in silviculture consist of plantation trees, such as eucalyptus, poplar, and pine.

Lignin Alteration

Several companies in the pulp and paper industry are interested in utilizing GM technology to alter the lignin content of plantation trees. It is estimated that reducing lignin in plantation trees by genetic modification could reduce pulping costs by up to $15 per cubic metre. Lignin removal from wood fibres conventionally relies on costly and environmentally hazardous chemicals. By developing low-lignin GM trees it is hoped that pulping and bleaching processes will require fewer inputs, therefore, mills supplied by low-lignin GM trees may have a reduced impact on their surrounding ecosystems and communities. However, it is argued that reductions in lignin may compromise the structural integrity of the plant, thereby making it more susceptible to pathogens and disease, which could necessitate pesticide use exceeding that on traditional plantations.

Frost Tolerance

Genetic modification can allow trees to cope with abiotic stresses such that their geographic range is broadened. Freeze-tolerant GM eucalyptus trees for use in southern US plantations are currently being tested in open air sites with such an objective in mind. ArborGen, a tree biotechnology company and joint venture of pulp and paper firms Rubicon (New Zealand), MeadWestvaco (US) and International Paper (US) is leading this research. Until now the cultivation of eucalyptus has only been possible on the southern tip of Florida, freeze-tolerance would substantially extend the cultivation range northwards.

Accelerated Growth

In Brazil, field trials of fast growing GM eucalyptus are currently underway, they are set to conclude in 2015-2016 with commercialization to result. *FuturaGene*, a biotechnology company owned by Suzano, a Brazilian pulp and paper company, has been leading this research. Stanley Hirsch, chief executive of FuturaGene has stated: "Our trees grow faster and thicker. We are ahead of everyone. We have shown we can increase the yields and growth rates of trees more than anything grown by traditional breeding." The company is looking to reduce harvest cycles from 7 to 5.5 years with 20-30% more mass than conventional eucalyptus. There is concern that such objectives may further exacerbate the negative impacts of plantation forestry. Increased water and soil nutrient demand from faster growing species may lead to irrecoverable losses in site productivity and further impinge upon neighbouring communities and ecosystems.

Disease Resistance

Ecologically motivated research into genetic modification is underway. There are ongoing schemes that aim to foster disease resistance in trees such as the American chestnut and the English elm for the purpose of their reintroduction to the wild. Specific diseases have reduced the populations of these emblematic species to the extent that they are mostly lost in the wild. Genetic modification is being pursued concurrently with traditional breeding techniques in an attempt to endow these species with disease resistance.

Current Uses

Poplars in China

In 2002 China's State Forestry Administration approved GM poplar trees for commercial use. Subsequently 1.4 million Bt (insecticide) producing GM poplars were planted in China. They were planted both for their wood and as part of China's 'Green Wall' project, which aims to impede desertification. Reports indicate that the GM poplars have spread beyond the area of original planting and that contamination of native poplars with the Bt gene is occurring. There is concern with these developments, particularly because the pesticide producing trait may impart a positive selective advantage on the poplar, allowing it a high level of invasiveness.

Agrobacterium

Agrobacterium is a genus of Gram-negative bacteria established by H. J. Conn that uses horizontal gene transfer to cause tumors in plants. *Agrobacterium tumefaciens* is the most commonly studied species in this genus. *Agrobacterium* is well known for its ability to transfer DNA between itself and plants, and for this reason it has become an important tool for genetic engineering.

The *Agrobacterium* genus is quite heterogeneous. Recent taxonomic studies have reclassified all of the *Agrobacterium* species into new genera, such as *Ahrensia*, *Pseudorhodobacter*, *Ruegeria*, and *Stappia*, but most species have been controversially reclassified as *Rhizobium* species.

Plant Pathogen

The large growths on these roots are galls induced by *Agrobacterium* sp.

A. tumefaciens causes crown-gall disease in plants. The disease is characterised by a tumour-like growth or gall on the infected plant, often at the junction between the root and the shoot. Tumors are incited by the conjugative transfer of a DNA segment (T-DNA) from the bacterial tumour-inducing (Ti) plasmid. The closely related species, *A. rhizogenes*, induces root tumors, and carries the distinct Ri (root-inducing) plasmid. Although the taxonomy of *Agrobacterium* is currently under revision it can be generalised that 3 biovars exist within the genus, *A. tumefaciens*, *A. rhizogenes*, and *A. vitis*. Strains within *A. tumefaciens* and *A. rhizogenes* are known to be able to harbour either a Ti or Ri-plasmid, whilst strains of *A. vitis*, generally restricted to grapevines, can harbour a Ti-plasmid. Non-*Agrobacterium* strains have been isolated from environmental samples which harbour a Ri-plasmid whilst laboratory studies have shown that non-*Agrobacterium* strains can also harbour a Ti-plasmid. Some environmental strains of *Agrobacterium* possess neither a Ti nor Ri-plasmid. These strains are avirulent.

The plasmid T-DNA is integrated semi-randomly into the genome of the host cell, and the tumor morphology genes on the T-DNA are expressed, causing the formation of a gall. The T-DNA carries genes for the biosynthetic enzymes for the production of unusual amino acids, typically octopine or nopaline. It also carries genes for the biosynthesis of the plant hormones, auxin and cytokinins, and for the biosynthesis of opines, providing a carbon and nitrogen source for the bacteria that most other micro-organisms can't use, giving *Agrobacterium* a selective advantage. By altering the hormone balance in the plant cell, the division of those cells cannot be controlled by the plant, and tumors form. The ratio of auxin to cytokinin produced by the tumor genes determines the morphology of the tumor (root-like, disorganized or shoot-like).

Agrobacterium in Humans

Although generally seen as an infection in plants, *Agrobacterium* can be responsible for opportunistic infections in humans with weakened immune systems, but has not been shown to be a primary pathogen in otherwise healthy individuals. One of the earliest associations of human disease caused by *Agrobacterium radiobacter* was reported by Dr. J. R. Cain in Scotland (1988). A later study suggested that *Agrobacterium* attaches to and genetically transforms several types of human cells by integrating its T-DNA into the human cell genome. The study was conducted using cultured human tissue and did not draw any conclusions regarding related biological activity in nature.

Uses in Biotechnology

The ability of *Agrobacterium* to transfer genes to plants and fungi is used in biotechnology, in particular, genetic engineering for plant improvement. A modified Ti or Ri plasmid can be used. The plasmid is 'disarmed' by deletion of the tumor inducing genes; the only essential parts of the T-DNA are its two small (25 base pair) border repeats, at least one of which is needed for plant transformation. The genes to be introduced into the plant are cloned into a plant transformation vector that contains the T-DNA region of the disarmed plasmid, together with a selectable marker (such as antibiotic resistance) to enable selection for plants that have been successfully transformed. Plants are grown on media containing antibiotic following transformation, and those that do not have the T-DNA integrated into their genome will die. An alternative method is agroinfiltration.

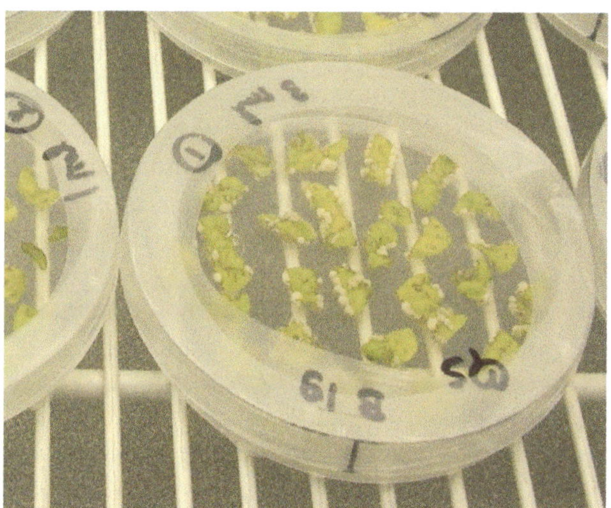

Plant (*S. chacoense*) transformed using *Agrobacterium*. Transformed cells start forming calluses on the side of the leaf pieces

Transformation with *Agrobacterium* can be achieved in two ways. Protoplasts or alternatively leaf-discs can be incubated with the *Agrobacterium* and whole plants regenerated using plant tissue culture. A common transformation protocol for *Arabidopsis* is the floral-dip method: inflorescence are dipped in a suspension of *Agrobacterium*, and the bacterium transforms the germline cells that make the female gametes. The seeds can then be screened for antibiotic resistance (or another marker of interest), and plants that have not integrated the plasmid DNA will die when exposed to the correct condition of antibiotic.

Agrobacterium does not infect all plant species, but there are several other effective techniques for plant transformation including the gene gun.

Agrobacterium is listed as being the vector of genetic material that was transferred to these USA GMOs:

- Soybean
- Cotton
- Corn
- Sugar Beet
- Alfalfa
- Wheat
- Rapeseed Oil (Canola)
- Creeping bentgrass (for animal feed)
- Rice (Golden Rice)

Genomics

The sequencing of the genomes of several species of *Agrobacterium* has permitted the study of the evolutionary history of these organisms and has provided information on the genes and systems involved in pathogenesis, biological control and symbiosis. One important finding is the possibility that chromosomes are evolving from plasmids in many of these bacteria. Another discovery is that the diverse chromosomal structures in this group appear to be capable of supporting both symbiotic and pathogenic lifestyles. The availability of the genome sequences of *Agrobacterium* species will continue to increase, resulting in substantial insights into the function and evolutionary history of this group of plant-associated microbes.

History

Marc Van Montagu and Jozef Schell at the University of Ghent (Belgium) discovered the gene transfer mechanism between *Agrobacterium* and plants, which resulted in the development of methods to alter *Agrobacterium* into an efficient delivery system for gene engineering in plants. A team of researchers led by Dr Mary-Dell Chilton were the first to demonstrate that the virulence genes could be removed without adversely affecting the ability of *Agrobacterium* to insert its own DNA into the plant genome (1983).

Tree Breeding

Tree breeding is the application of genetic, reproductive biology and economics principles to the genetic improvement and management of forest trees. In contrast to the selective breeding of livestock, arable crops, and horticultural flowers over the last few centuries, the breeding of trees, with the exception of fruit trees, is a relatively recent occurrence.

A typical forest tree breeding program starts with selection of superior phenotypes (plus trees) in a natural or planted forest. This application of mass selection improves the mean performance of the forest. Offspring is obtained from selected trees and grown in test plantations that act as genetic trials. Based on such tests the best genotypes among the parents can be selected. Selected trees are typically multiplied by either seeds or grafting and seed orchards are established when the preferred output is improved seed. Alternatively, the best genotypes can be directly propagated by cuttings or in-vitro methods and used directly in clonal plantations. The first system is frequently used in pines and other conifers, while the second is typical in some broadleaves (poplars, eucalypts and others). The objectives of a tree breeding program range from yield improvement and adaptation to particular conditions, to pest- and disease-resistance, wood properties, etc. Currently, tree breeding starting to take advantage of the fast development in plant genetics and genomics.

Optimisation

Tree breeders make efforts to get their operation efficient by optimising tree breeding. Scientists develop tools aimed at improvement of the efficiency of tree breeding programmes. Optimising can mean adapting strategies and methods to certain species, group of populations, structure of genetic variation and mode of inheritance of the important traits to obtain the highest benefit per unit of time. Optimising is usually carried out at the following levels: breeding strategy (appropriate intensity of breeding, breeding population structure and size, plan for maintenance of genetic diversity), breeding methods (mating type, testing and selection methods, testing population size and time) and deployment methods of the genetically improved material (seed orchards and clonal forestry: genetic contribution, size). Computer simulators are frequently used: stochastic – based on defined and random algorithms; and deterministic – based on defined algorithms.

Selection strategies have been compared for annual progress in long-term breeding at a given annual cost considering genetic gain, gene diversity, cost components, and time components. For Norway spruce it seems favourable to clone full sib families and then select based on clonal performance while for Scots pine a two-stage strategy seems best, first phenotypic pre-selection and then progeny-testing the selections.

Tree Improvement

A genetically variable population and a method of selecting genetically superior individuals provide the basis for tree improvement by breeding. In essence, a tree improvement program sets out to isolate and evaluate the genetic component of variation in one or more characters of interest. In the simplest procedure, cycles of selection reduce the available population in a particular direction to enhance desirable traits, then breeding from selections to expand the population with improved characteristics. Breeding strategies vary with species and objectives, but all use mating designs to generate information and new material. Choice of a suitable breeding strategy and mating design is a key decision in any breeding program. Kiss (1986) used a 2-level design in British Columbia to study variation within and between separate populations of white spruce, both within British Columbia and from eastern North America.

The breeding program for white spruce initiated in 1986 by the Canadian Forestry Service in the Maritimes employed 2 kinds of mating: polycross, to test clones for general combining ability; and pair-mating, to generate material for second generation selections (Fowler et al. 1988).

Newton's (2003) systematic review of yield responses of white spruce and 3 other North American conifers to forest tree improvement practices indicated that correct provenance-progeny selection could yield juvenile height growth gains of about 12% at 20 years for white spruce, and a corresponding merchantable productivity (mean annual merchantable volume increment) gain of 26% at 50 years for plantations established at nominal initial densities on medium-to-good quality sites. Also, preliminary estimates derived from individual case studies indicated that first generational selection strategies for white spruce could increase merchantable productivity by approximately 20% at 45 years.

Genetically Modified Crops

Genetically modified crops (GMCs, GM crops, or biotech crops) are plants used in agriculture, the DNA of which has been modified using genetic engineering techniques. In most cases, the aim is to introduce a new trait to the plant which does not occur naturally in the species. Examples in food crops include resistance to certain pests, diseases, or environmental conditions, reduction of spoilage, or resistance to chemical treatments (e.g. resistance to a herbicide), or improving the nutrient profile of the crop. Examples in non-food crops include production of pharmaceutical agents, biofuels, and other industrially useful goods, as well as for bioremediation.

Farmers have widely adopted GM technology. Between 1996 and 2015, the total surface area of land cultivated with GM crops increased by a factor of 100, from 17,000 km^2 (4.2 million acres) to 1,797,000 km^2 (444 million acres). 10% of the world's arable land was planted with GM crops in 2010. In the US, by 2014, 94% of the planted area of soybeans, 96% of cotton and 93% of corn were genetically modified varieties. Use of GM crops expanded rapidly in developing countries, with about 18 million farmers growing 54% of worldwide GM crops by 2013. A 2014 meta-analysis concluded that GM technology adoption had reduced chemical pesticide use by 37%, increased crop yields by 22%, and increased farmer profits by 68%. This reduction in pesticide use has been ecologically beneficial, but benefits may be reduced by overuse. Yield gains and pesticide reductions are larger for insect-resistant crops than for herbicide-tolerant crops. Yield and profit gains are higher in developing countries than in developed countries.

There is a scientific consensus that currently available food derived from GM crops poses no greater risk to human health than conventional food, but that each GM food needs to be tested on a case-by-case basis before introduction. Nonetheless, members of the public are much less likely than scientists to perceive GM foods as safe. The legal and regulatory status of GM foods varies by country, with some nations banning or restricting them, and others permitting them with widely differing degrees of regulation.

However, opponents have objected to GM crops on several grounds, including environmental concerns, whether food produced from GM crops is safe, whether GM crops are needed to address the world's food needs, and concerns raised by the fact these organisms are subject to intellectual property law.

Gene Transfer in Nature and Traditional Agriculture

DNA transfers naturally between organisms. Several natural mechanisms allow gene flow across species. These occur in nature on a large scale – for example, it is one mechanism for the development of antibiotic resistance in bacteria. This is facilitated by transposons, retrotransposons, proviruses and other mobile genetic elements that naturally translocate DNA to new loci in a genome. Movement occurs over an evolutionary time scale.

The introduction of foreign germplasm into crops has been achieved by traditional crop breeders by overcoming species barriers. A hybrid cereal grain was created in 1875, by crossing wheat and rye. Since then important traits including dwarfing genes and rust resistance have been introduced. Plant tissue culture and deliberate mutations have enabled humans to alter the makeup of plant genomes.

History

The first genetically modified crop plant was produced in 1982, an antibiotic-resistant tobacco plant. The first field trials occurred in France and the USA in 1986, when tobacco plants were engineered for herbicide resistance. In 1987, Plant Genetic Systems (Ghent, Belgium), founded by Marc Van Montagu and Jeff Schell, was the first company to genetically engineer insect-resistant (tobacco) plants by incorporating genes that produced insecticidal proteins from *Bacillus thuringiensis* (Bt).

The People's Republic of China was the first country to allow commercialized transgenic plants, introducing a virus-resistant tobacco in 1992, which was withdrawn in 1997. The first genetically modified crop approved for sale in the U.S., in 1994, was the *FlavrSavr* tomato. It had a longer shelf life, because it took longer to soften after ripening. In 1994, the European Union approved tobacco engineered to be resistant to the herbicide bromoxynil, making it the first commercially genetically engineered crop marketed in Europe.

In 1995, Bt Potato was approved by the US Environmental Protection Agency, making it the country's first pesticide producing crop. In 1995 canola with modified oil composition (Calgene), Bt maize (Ciba-Geigy), bromoxynil-resistant cotton (Calgene), Bt cotton (Monsanto), glyphosate-resistant soybeans (Monsanto), virus-resistant squash (Asgrow), and additional delayed ripening tomatoes (DNAP, Zeneca/Peto, and Monsanto) were approved. As of mid-1996, a total of 35 approvals had been granted to commercially grow 8 transgenic crops and one flower crop (carnation), with 8 different traits in 6 countries plus the EU. In 2000, Vitamin A-enriched golden rice was developed, though as of 2016 it was not yet in commercial production. In 2013 the leaders of the three research teams that first applied genetic engineering to crops, Robert Fraley, Marc Van Montagu and Mary-Dell Chilton were awarded the World Food Prize for improving the "quality, quantity or availability" of food in the world.

Methods

Genetically engineered crops have genes added or removed using genetic engineering techniques, originally including gene guns, electroporation, microinjection and agrobacterium. More recently, CRISPR and TALEN offered much more precise and convenient editing techniques.

Gene guns (a.k.a. biolistic) "shoot" (direct high energy particles or radiations against) target genes into plant cells. It is the most common method. DNA is bound to tiny particles of gold or tungsten which are subsequently shot into plant tissue or single plant cells under high pressure. The accelerated particles penetrate both the cell wall and membranes. The DNA separates from the metal and is integrated into plant DNA inside the nucleus. This method has been applied successfully for many cultivated crops, especially monocots like wheat or maize, for which transformation using *Agrobacterium tumefaciens* has been less successful. The major disadvantage of this procedure is that serious damage can be done to the cellular tissue.

Agrobacterium tumefaciens-mediated transformation is another common technique. Agrobacteria are natural plant parasites, and their natural ability to transfer genes provides another engineering method. To create a suitable environment for themselves, these Agrobacteria insert their genes into plant hosts, resulting in a proliferation of modified plant cells near the soil level (crown gall). The genetic information for tumour growth is encoded on a mobile, circular DNA fragment (plasmid). When Agrobacterium infects a plant, it transfers this T-DNA to a random site in the plant genome. When used in genetic engineering the bacterial T-DNA is removed from the bacterial plasmid and replaced with the desired foreign gene. The bacterium is a vector, enabling transportation of foreign genes into plants. This method works especially well for dicotyledonous plants like potatoes, tomatoes, and tobacco. Agrobacteria infection is less successful in crops like wheat and maize.

Electroporation is used when the plant tissue does not contain cell walls. In this technique, "DNA enters the plant cells through miniature pores which are temporarily caused by electric pulses."

Microinjection directly injects the gene into the DNA.

Plant scientists, backed by results of modern comprehensive profiling of crop composition, point out that crops modified using GM techniques are less likely to have unintended changes than are conventionally bred crops.

In research tobacco and *Arabidopsis thaliana* are the most frequently modified plants, due to well-developed transformation methods, easy propagation and well studied genomes. They serve as model organisms for other plant species.

Introducing new genes into plants requires a promoter specific to the area where the gene is to be expressed. For instance, to express a gene only in rice grains and not in leaves, an endosperm-specific promoter is used. The codons of the gene must be optimized for the organism due to codon usage bias. Transgenic gene products should be able to be denatured by heat so that they are destroyed during cooking.

Types of Modifications

Transgenic

Transgenic plants have genes inserted into them that are derived from another species. The inserted genes can come from species within the same kingdom (plant to plant) or between kingdoms (for example, bacteria to plant). In many cases the inserted DNA has to be modified slightly in order to correctly and efficiently express in the host organism. Transgenic plants are used to express proteins like the cry toxins from *B. thuringiensis*, herbicide resistant genes, antibodies and antigens for vaccinations A study led by the European Food Safety Authority (EFSA) found also viral genes in transgenic plants.

Transgenic maize containing a gene from the bacteria *Bacillus thuringiensis*

Transgenic carrots have been used to produce the drug Taliglucerase alfa which is used to treat Gaucher's disease. In the laboratory, transgenic plants have been modified to increase photosynthesis (currently about 2% at most plants versus the theoretic potential of 9–10%). This is possible by changing the rubisco enzyme (i.e. changing C3 plants into C4 plants), by placing the rubisco in a carboxysome, by adding CO_2 pumps in the cell wall, by changing the leaf form/size. Plants have been engineered to exhibit bioluminescence that may become a sustainable alternative to electric lighting.

Cisgenic

Cisgenic plants are made using genes found within the same species or a closely related one, where conventional plant breeding can occur. Some breeders and scientists argue that cisgenic modification is useful for plants that are difficult to crossbreed by conventional means (such as potatoes), and that plants in the cisgenic category should not require the same regulatory scrutiny as transgenics.

Subgenic

Genetically modified plants can also be developed using gene knockdown to alter the genetic make-up of a plant without incorporating genes from other plants. In 2014, Chinese researcher Gao Caixia filed patents on the creation of a strain of wheat that is resistant to powdery mildew. The strain lacks genes that encode proteins that repress defenses against the mildew. The researchers deleted all three copies of the genes from wheat's hexaploid genome. Gao used the TALENs and CRISPR gene editing tools without adding or changing any other genes. No field trials were immediately planned. The CRISPR technique has also been used to modify white button mushrooms (*Agaricus bisporus*).

Economics

GM food's economic value to farmers is one of its major benefits, including in developing nations. A 2010 study found that Bt corn provided economic benefits of $6.9 billion over the previous 14 years in five Midwestern states. The majority ($4.3 billion) accrued to farmers producing non-Bt corn. This was attributed to European corn borer populations reduced by exposure to Bt corn, leaving fewer to attack conventional corn nearby. Agriculture economists calculated that "world surplus [increased by] $240.3 million for 1996. Of this total, the largest share (59%) went to U.S. farmers. Seed company Monsanto received the next largest share (21%), followed by US consumers (9%), the rest of the world (6%), and the germplasm supplier, Delta & Pine Land Company of Mississippi (5%)."

According to the International Service for the Acquisition of Agri-biotech Applications (ISAAA), in 2014 approximately 18 million farmers grew biotech crops in 28 countries; about 94% of the farmers were resource-poor in developing countries. 53% of the global biotech crop area of 181.5 million hectares was grown in 20 developing countries. PG Economics comprehensive 2012 study concluded that GM crops increased farm incomes worldwide by $14 billion in 2010, with over half this total going to farmers in developing countries.

Critics challenged the claimed benefits to farmers over the prevalence of biased observers and by the absence of randomized controlled trials. The main Bt crop grown by small farmers in developing countries is cotton. A 2006 review of Bt cotton findings by agricultural economists concluded, "the overall balance sheet, though promising, is mixed. Economic returns are highly variable over years, farm type, and geographical location".

In 2013 the European Academies Science Advisory Council (EASAC) asked the EU to allow the development of agricultural GM technologies to enable more sustainable agriculture, by employing fewer land, water and nutrient resources. EASAC also criticizes the EU's "timeconsuming and expensive regulatory framework" and said that the EU had fallen behind in the adoption of GM technologies.

Participants in agriculture business markets include seed companies, agrochemical companies, distributors, farmers, grain elevators and universities that develop new crops/traits and whose agricultural extensions advise farmers on best practices. According to a 2012 review based on data from the late 1990s and early 2000s, much of the GM crop grown each year is used for livestock feed and increased demand for meat leads to increased demand for GM feedcrops. Feed grain usage as a percentage of total crop production is 70% for corn and more than 90% of oil seed meals such as soybeans. About 65 million metric tons of GM corn grains and about 70 million metric tons of soybean meals derived from GM soybean become feed.

In 2014 the global value of biotech seed was US$15.7 billion; US$11.3 billion (72%) was in industrial countries and US$4.4 billion (28%) was in the developing countries. In 2009, Monsanto had $7.3 billion in sales of seeds and from licensing its technology; DuPont, through its Pioneer subsidiary, was the next biggest company in that market. As of 2009, the overall Roundup line of products including the GM seeds represented about 50% of Monsanto's business.

Some patents on GM traits have expired, allowing the legal development of generic strains that include these traits. For example, generic glyphosate-tolerant GM soybean is now available. Another impact is that traits developed by one vendor can be added to another vendor's proprietary strains, potentially increasing product choice and competition. The patent on the first type of *Roundup Ready* crop that Monsanto produced (soybeans) expired in 2014 and the first harvest of off-patent soybeans occurs in the spring of 2015. Monsanto has broadly licensed the patent to other seed companies that include the glyphosate resistance trait in their seed products. About 150 companies have licensed the technology, including Syngenta and DuPont Pioneer.

Yield

In 2014, the largest review yet concluded that GM crops' effects on farming were positive. The meta-analysis considered all published English-language examinations of the agronomic and economic impacts between 1995 and March 2014 for three major GM crops: soybean, maize, and cotton. The study found that herbicide-tolerant crops have lower production costs, while for insect-resistant crops the reduced pesticide use was offset by higher seed prices, leaving overall production costs about the same.

Yields increased 9% for herbicide tolerance and 25% for insect resistant varieties. Farmers who adopted GM crops made 69% higher profits than those who did not. The review found that GM crops help farmers in developing countries, increasing yields by 14 percentage points.

The researchers considered some studies that were not peer-reviewed, and a few that did not report sample sizes. They attempted to correct for publication bias, by considering sources beyond academic journals. The large data set allowed the study to control for potentially confounding variables such as fertiliser use. Separately, they concluded that the funding source did not influence study results.

Traits

GM crops grown today, or under development, have been modified with various traits. These traits include improved shelf life, disease resistance, stress resistance, herbicide resistance, pest resistance, production of useful goods such as biofuel or drugs, and ability to absorb toxins and for use in bioremediation of pollution.

Recently, research and development has been targeted to enhancement of crops that are locally important in developing countries, such as insect-resistant cowpea for Africa and insect-resistant brinjal (eggplant).

Lifetime

The first genetically modified crop approved for sale in the U.S. was the *FlavrSavr* tomato, which had a longer shelf life. It is no longer on the market.

In November 2014, the USDA approved a GM potato that prevents bruising.

In February 2015 Arctic Apples were approved by the USDA, becoming the first genetically modified apple approved for US sale. Gene silencing was used to reduce the expression of polyphenol oxidase (PPO), thus preventing enzymatic browning of the fruit after it has been sliced open. The trait was added to Granny Smith and Golden Delicious varieties. The trait includes a bacterial antibiotic resistance gene that provides resistance to the antibiotic kanamycin. The genetic engineering involved cultivation in the presence of kanamycin, which allowed only resistant cultivars to survive. Humans consuming apples do not acquire kanamycin resistance, per arcticapple.com. The FDA approved the apples in March 2015.

Nutrition

Edible Oils

Some GM soybeans offer improved oil profiles for processing or healthier eating. Camelina sativa has been modified to produce plants that accumulate high levels of oils similar to fish oils.

Vitamin Enrichment

Golden rice, developed by the International Rice Research Institute (IRRI), provides greater amounts of Vitamin A targeted at reducing Vitamin A deficiency. As of January 2016, golden rice has not yet been grown commercially in any country.

Researchers vitamin-enriched corn derived from South African white corn variety M37W, producing a 169-fold increase in Vitamin A, 6-fold increase in Vitamin C and doubled concentrations of folate. Modified Cavendish bananas express 10-fold the amount of Vitamin A as unmodified varieties.

Toxin Reduction

A genetically modified cassava under development offers lower cyanogen glucosides and enhanced protein and other nutrients (called BioCassava).

In November 2014, the USDA approved a potato, developed by J.R. Simplot Company, that prevents bruising and produces less acrylamide when fried. The modifications prevent natural, harmful proteins from being made via RNA interference. They do not employ genes from non-potato species. The trait was added to the Russet Burbank, Ranger Russet and Atlantic varieties.

Stress Resistance

Plants engineered to tolerate non-biological stressors such as drought, frost, high soil salinity, and nitrogen starvation were in development. In 2011, Monsanto's DroughtGard maize became the first drought-resistant GM crop to receive US marketing approval.

Herbicides

Glyphosate

As of 1999 the most prevalent GM trait was glyphosate-resistance. Glyphosate, (the active ingredient in Roundup and other herbicide products) kills plants by interfering with the shikimate pathway in plants, which is essential for the synthesis of the aromatic amino acids phenylalanine, tyrosine and tryptophan. The shikimate pathway is not present in animals, which instead obtain aromatic amino acids from their diet. More specifically, glyphosate inhibits the enzyme 5-enolpyruvylshikimate-3-phosphate synthase (EPSPS).

This trait was developed because the herbicides used on grain and grass crops at the time were highly toxic and not effective against narrow-leaved weeds. Thus, developing crops that could withstand spraying with glyphosate would both reduce environmental and health risks, and give an agricultural edge to the farmer.

Some micro-organisms have a version of EPSPS that is resistant to glyphosate inhibition. One of these was isolated from an *Agrobacterium* strain CP4 (CP4 EPSPS) that was resistant to glyphosate. The CP4 EPSPS gene was engineered for plant expression by fusing the 5' end of the gene to a chloroplast transit peptide derived from the petunia EPSPS. This transit peptide was used because it had shown previously an ability to deliver bacterial EPSPS to the chloroplasts of other plants. This CP4 EPSPS gene was cloned and transfected into soybeans.

The plasmid used to move the gene into soybeans was PV-GMGTO4. It contained three bacterial genes, two CP4 EPSPS genes, and a gene encoding beta-glucuronidase (GUS) from *Escherichia coli* as a marker. The DNA was injected into the soybeans using the particle acceleration method. Soybean cultivar A5403 was used for the transformation.

Bromoxynil

Tobacco plants have been engineered to be resistant to the herbicide bromoxynil.

Glufosinate

Crops have been commercialized that are resistant to the herbicide glufosinate, as well. Crops engineered for resistance to multiple herbicides to allow farmers to use a mixed group of two, three, or four different chemicals are under development to combat growing herbicide resistance.

2,4-D

In October 2014 the US EPA registered Dow's Enlist Duo maize, which is genetically modified to be resistant to both glyphosate and 2,4-D, in six states. Inserting a bacterial aryloxyalkanoate dioxygenase gene, *aad1* makes the corn resistant to 2,4-D. The USDA had approved maize and soybeans with the mutation in September 2014.

Dicamba

Monsanto has requested approval for a stacked strain that is tolerant of both glyphosate and dicamba.

Pest Resistance

Insects

Tobacco, corn, rice and many other crops have been engineered to express genes encoding for insecticidal proteins from Bacillus thuringiensis (Bt). Papaya, potatoes, and squash have been engineered to resist viral pathogens such as cucumber mosaic virus which, despite its name, infects a wide variety of plants. The introduction of Bt crops during the period between 1996 and 2005 has been estimated to have reduced the total volume of insecticide active ingredient use in the United States by over 100 thousand tons. This represents a 19.4% reduction in insecticide use.

In the late 1990s, a genetically modified potato that was resistant to the Colorado potato beetle was withdrawn because major buyers rejected it, fearing consumer opposition.

Viruses

Virus resistant papaya were developed In response to a papaya ringspot virus (PRV) outbreak in Hawaii in the late 1990s. They incorporate PRV DNA. By 2010, 80% of Hawaiian papaya plants were genetically modified.

Potatoes were engineered for resistance to potato leaf roll virus and Potato virus Y in 1998. Poor sales led to their market withdrawal after three years.

Yellow squash that were resistant to at first two, then three viruses were developed, beginning in the 1990s. The viruses are watermelon, cucumber and zucchini/courgette yellow mosaic. Squash was the second GM crop to be approved by US regulators. The trait was later added to zucchini.

Many strains of corn have been developed in recent years to combat the spread of Maize dwarf mosaic virus, a costly virus that causes stunted growth which is carried in Johnson grass and spread by aphid insect vectors. These strands are commercially available although the resistance is not standard among GM corn variants.

By-products

Drugs

In 2012, the FDA approved the first plant-produced pharmaceutical, a treatment for Gaucher's Disease. Tobacco plants have been modified to produce therapeutic antibodies.

Biofuel

Algae is under development for use in biofuels. Researchers in Singapore were working on GM jatropha for biofuel production. Syngenta has USDA approval to market a maize trademarked Enogen that has been genetically modified to convert its starch to sugar for ethanol. In 2013, the Flemish Institute for Biotechnology was investigating poplar trees genetically engineered to contain less lignin to ease conversion into ethanol. Lignin is the critical limiting factor when using wood to make bio-ethanol because lignin limits the accessibility of cellulose microfibrils to depolymerization by enzymes.

Materials

Companies and labs are working on plants that can be used to make bioplastics. Potatoes that produce industrially useful starches have been developed as well. Oilseed can be modified to produce fatty acids for detergents, substitute fuels and petrochemicals.

Bioremediation

Scientists at the University of York developed a weed (*Arabidopsis thaliana*) that contains genes from bacteria that can clean TNT and RDX-explosive soil contaminants. 16 million hectares in the USA (1.5% of the total surface) are estimated to be contaminated with TNT and RDX. However *A. thaliana* was not tough enough for use on military test grounds.

Genetically modified plants have been used for bioremediation of contaminated soils. Mercury, selenium and organic pollutants such as polychlorinated biphenyls (PCBs).

Marine environments are especially vulnerable since pollution such as oil spills are not containable. In addition to anthropogenic pollution, millions of tons of petroleum annually enter the marine environment from natural seepages. Despite its toxicity, a considerable fraction of petroleum oil entering marine systems is eliminated by the hydrocarbon-degrading activities of microbial communities. Particularly successful is a recently discovered group of specialists, the so-called hydrocarbonoclastic bacteria (HCCB) that may offer useful genes.

Asexual Reproduction

Crops such as maize reproduce sexually each year. This randomizes which genes get propagated to the next generation, meaning that desirable traits can be lost. To maintain a high-quality crop, some farmers purchase seeds every year. Typically, the seed company maintains two inbred varieties, and crosses them into a hybrid strain that is then sold. Related plants like sorghum and gamma grass are able to perform apomixis, a form of asexual reproduction that keeps the plant's DNA intact. This trait is apparently controlled by a single dominant gene, but traditional breeding has been unsuccessful in creating asexually-reproducing maize. Genetic engineering offers another route to this goal. Successful modification would allow farmers to replant harvested seeds that retain desirable traits, rather than relying on purchased seed.

Crops

Herbicide Tolerance

GMO	Use	Countries approved in	First approved	Notes
Alfalfa	Animal feed	USA	2005	Approval withdrawn in 2007 and then re-approved in 2011
Canola	Cooking oil	Australia	2003	
	Margarine	Canada	1995	
	Emulsifiers in packaged foods	USA	1995	

GMO	Use	Countries approved in	First approved	Notes
Cotton	Fiber Cottonseed oil Animal feed	Argentina	2001	
		Australia	2002	
		Brazil	2008	
		Columbia	2004	
		Costa Rica	2008	
		Mexico	2000	
		Paraguay	2013	
		South Africa	2000	
		USA	1994	
Maize	Animal feed high-fructose corn syrup corn starch	Argentina	1998	
		Brazil	2007	
		Canada	1996	
		Colombia	2007	
		Cuba	2011	
		European Union	1998	Grown in Portugal, Spain, Czech Republic, Slovakia and Romania
		Honduras	2001	
		Paraguay	2012	
		Philippines	2002	
		South Africa	2002	
		USA	1995	
		Uruguay	2003	
Soybean	Animal feed Soybean oil	Argentina	1996	
		Bolivia	2005	
		Brazil	1998	
		Canada	1995	
		Chile	2007	
		Costa Rica	2001	
		Mexico	1996	
		Paraguay	2004	
		South Africa	2001	
		USA	1993	
		Uruguay	1996	
Sugar Beet	Food	Canada	2001	
		USA	1998	Commercialised 2007, production blocked 2010, resumed 2011.

Insect Resistance

GMO	Use	Countries approved in	First approved	Notes
Cotton	Fiber	Argentina	1998	
		Australia	2003	
		Brazil	2005	
		Burkina Faso	2009	
		China	1997	
		Colombia	2003	
		Costa Rica	2008	
		India	2002	Largest producer of Bt cotton
		Mexico	1996	
		Myanmar	2006[N 1]	
		Pakistan	2010[N 1]	
		Paraguay	2007	
		South Africa	1997	
		Sudan	2012	
		USA	1995	
Eggplant	Food	Bangladesh	2013	12 ha planted on 120 farms in 2014
Maize	Animal feed	Argentina	1998	
		Brazil	2005	
		Columbia	2003	
		Mexico	1996	Centre of origin for maize
		Paraguay	2007	
		Philippines	2002	
		South Africa	1997	
		Uruguay	2003	
		USA	1995	
Poplar	Tree	China	1998	543 ha of bt poplar planted in 2014

Other Modified Traits

GMO	Use	Trait	Countries approved in	First approved	Notes
Canola	Cooking oil Margarine Emulsifiers in packaged foods	High laurate canola	Canada	1996	
			USA	1994	
		Phytase production	USA	1998	

GMO	Use	Trait	Countries approved in	First approved	Notes
Carnation	Ornamental	Delayed senescence	Australia	1995	
			Norway	1998	
		Modified flower colour	Australia	1995	
			Columbia	2000	In 2014 4 ha were grown in greenhouses for export
			European Union	1998	Two events expired 2008, another approved 2007
			Japan	2004	
			Malaysia	2012	For ornamental purposes
			Norway	1997	
Maize	Animal feed high-fructose corn syrup corn starch	Increased lysine	Canada	2006	
			USA	2006	
		Drought tolerance	Canada	2010	
			USA	2011	
Papaya	Food	Virus resistance	China	2006	
			USA	1996	Mostly grown in Hawaii
Petunia	Ornamental	Modified flower colour		1998[N 1]	
Potato	Food	Virus resistance	Canada	1999	
			USA	1997	
	Industrial	Modified starch	USA	2014	
Rose	Ornamental	Modified flower colour	Australia	2009	Surrendered renewal
			Colombia	2010[N 2]	Greenhouse cultivation for export only.
			Japan	2008	
			USA	2011	
Soybean	Animal feed Soybean oil	Increased oleic acid production	Argentina	2015	
			Canada	2000	
			USA	1997	
		Stearidonic acid production	Canada	2011	
			USA	2011	
Squash	Food	Virus resistance	USA	1994	
Sugar Cane	Food	Drought tolerance	Indonesia	2013	Environmental certificate only
Tobacco	Cigarettes	Nicotine reduction	USA	2002	

Development

The number of USDA-approved field releases for testing grew from 4 in 1985 to 1,194 in 2002 and averaged around 800 per year thereafter. The number of sites per release and the number of gene constructs (ways that the gene of interest is packaged together with other elements)—have rapidly increased since 2005. Releases with agronomic properties (such as drought resistance) jumped from 1,043 in 2005 to 5,190 in 2013. As of September 2013, about 7,800 releases had been approved for corn, more than 2,200 for soybeans, more than 1,100 for cotton, and about 900 for potatoes. Releases were approved for herbicide tolerance (6,772 releases), insect resistance (4,809), product quality such as flavor or nutrition (4,896), agronomic properties like drought resistance (5,190), and virus/fungal resistance (2,616). The institutions with the most authorized field releases include Monsanto with 6,782, Pioneer/DuPont with 1,405, Syngenta with 565, and USDA's Agricultural Research Service with 370. As of September 2013 USDA had received proposals for releasing GM rice, squash, plum, rose, tobacco, flax and chicory.

Farming Practices

Resistance

Bt Resistance

Constant exposure to a toxin creates evolutionary pressure for pests resistant to that toxin. Over-reliance on glyphosate and a reduction in the diversity of weed management practices allowed the spread of glyphosate resistance in 14 weed species/biotypes in the US.

One method of reducing resistance is the creation of refuges to allow nonresistant organisms to survive and maintain a susceptible population.

To reduce resistance to Bt crops, the 1996 commercialization of transgenic cotton and maize came with a management strategy to prevent insects from becoming resistant. Insect resistance management plans are mandatory for Bt crops. The aim is to encourage a large population of pests so that any (recessive) resistance genes are diluted within the population. Resistance lowers evolutionary fitness in the absence of the stressor (Bt). In refuges, non-resistant strains outcompete resistant ones.

With sufficiently high levels of transgene expression, nearly all of the heterozygotes (S/s), i.e., the largest segment of the pest population carrying a resistance allele, will be killed before maturation, thus preventing transmission of the resistance gene to their progeny. Refuges (i. e., fields of non-transgenic plants) adjacent to transgenic fields increases the likelihood that homozygous resistant (s/s) individuals and any surviving heterozygotes will mate with susceptible (S/S) individuals from the refuge, instead of with other individuals carrying the resistance allele. As a result, the resistance gene frequency in the population remains lower.

Complicating factors can affect the success of the high-dose/refuge strategy. For example, if the temperature is not ideal, thermal stress can lower Bt toxin production and leave the plant more susceptible. More importantly, reduced late-season expression has been documented, possibly resulting from DNA methylation of the promoter. The success of the high-dose/refuge strategy has successfully maintained the value of Bt crops. This success has depended on factors independent of management strategy, including low initial resistance allele frequencies, fitness costs associated with resistance, and the abundance of non-Bt host plants outside the refuges.

Companies that produce Bt seed are introducing strains with multiple Bt proteins. Monsanto did this with Bt cotton in India, where the product was rapidly adopted. Monsanto has also; in an attempt to simplify the process of implementing refuges in fields to comply with Insect Resistance Management(IRM) policies and prevent irresponsible planting practices; begun marketing seed bags with a set proportion of refuge (non-transgenic) seeds mixed in with the Bt seeds being sold. Coined "Refuge-In-a-Bag" (RIB), this practice is intended to increase farmer compliance with refuge requirements and reduce additional labor needed at planting from having separate Bt and refuge seed bags on hand. This strategy is likely to reduce the likelihood of Bt-resistance occurring for corn rootworm, but may increase the risk of resistance for lepidopteran corn pests, such as European corn borer. Increased concerns for resistance with seed mixtures include partially resistant larvae on a Bt plant being able to move to a susceptible plant to survive or cross pollination of refuge pollen on to Bt plants that can lower the amount of Bt expressed in kernels for ear feeding insects.

Herbicide Resistance

Best management practices (BMPs) to control weeds may help delay resistance. BMPs include applying multiple herbicides with different modes of action, rotating crops, planting weed-free seed, scouting fields routinely, cleaning equipment to reduce the transmission of weeds to other fields, and maintaining field borders. The most widely planted GMOs are designed to tolerate herbicides. By 2006 some weed populations had evolved to tolerate some of the same herbicides. Palmer amaranth is a weed that competes with cotton. A native of the southwestern US, it traveled east and was first found resistant to glyphosate in 2006, less than 10 years after GM cotton was introduced.

Plant Protection

Farmers generally use less insecticide when they plant Bt-resistant crops. Insecticide use on corn farms declined from 0.21 pound per planted acre in 1995 to 0.02 pound in 2010. This is consistent with the decline in European corn borer populations as a direct result of Bt corn and cotton. The establishment of minimum refuge requirements helped delay the evolution of Bt resistance. However resistance appears to be developing to some Bt traits in some areas.

Tillage

By leaving at least 30% of crop residue on the soil surface from harvest through planting, conservation tillage reduces soil erosion from wind and water, increases water retention, and reduces soil degradation as well as water and chemical runoff. In addition, conservation tillage reduces the carbon footprint of agriculture. A 2014 review covering 12 states from 1996 to 2006, found that a 1% increase in herbicde-tolerant (HT) soybean adoption leads to a 0.21% increase in conservation tillage and a 0.3% decrease in quality-adjusted herbicide use.

Regulation

The regulation of genetic engineering concerns the approaches taken by governments to assess and manage the risks associated with the development and release of genetically modified crops. There are differences in the regulation of GM crops between countries, with some of the most marked differences occurring between the USA and Europe. Regulation varies in a given country depending on the intended use of each product. For example, a crop not intended for food use is generally not reviewed by authorities responsible for food safety.

Production

In 2013, GM crops were planted in 27 countries; 19 were developing countries and 8 were developed countries. 2013 was the second year in which developing countries grew a majority (54%) of the total GM harvest. 18 million farmers grew GM crops; around 90% were small-holding farmers in developing countries.

Country	2013– GM planted area (million hectares)	Biotech crops
USA	70.1	Maize, Soybean, Cotton, Canola, Sugarbeet, Alfalfa, Papaya, Squash
Brazil	40.3	Soybean, Maize, Cotton
Argentina	24.4	Soybean, Maize, Cotton
India	11.0	Cotton
Canada	10.8	Canola, Maize, Soybean, Sugarbeet
Total	175.2	----

The United States Department of Agriculture (USDA) reports every year on the total area of GMO varieties planted in the United States. According to National Agricultural Statistics Service, the states published in these tables represent 81–86 percent of all corn planted area, 88–90 percent of all soybean planted area, and 81–93 percent of all upland cotton planted area (depending on the year).

Global estimates are produced by the International Service for the Acquisition of Agri-biotech Applications (ISAAA) and can be found in their annual reports, "Global Status of Commercialized Transgenic Crops".

Farmers have widely adopted GM technology. Between 1996 and 2013, the total surface area of land cultivated with GM crops increased by a factor of 100, from 17,000 square kilometers (4,200,000 acres) to 1,750,000 km^2 (432 million acres). 10% of the world's arable land was planted with GM crops in 2010. As of 2011, 11 different transgenic crops were grown commercially on 395 million acres (160 million hectares) in 29 countries such as the USA, Brazil, Argentina, India, Canada, China, Paraguay, Pakistan, South Africa, Uruguay, Bolivia, Australia, Philippines, Myanmar, Burkina Faso, Mexico and Spain. One of the key reasons for this widespread adoption is the perceived economic benefit the technology brings to farmers. For example, the system of planting glyphosate-resistant seed and then applying glyphosate once plants emerged provided farmers with the opportunity to dramatically increase the yield from a given plot of land, since this allowed them to plant rows closer together. Without it, farmers had to plant rows far enough apart to control post-emergent weeds with mechanical tillage. Likewise, using Bt seeds means that farmers do not have to purchase insecticides, and then invest time, fuel, and equipment in applying them. However critics have disputed whether yields are higher and whether chemical use is less, with GM crops.

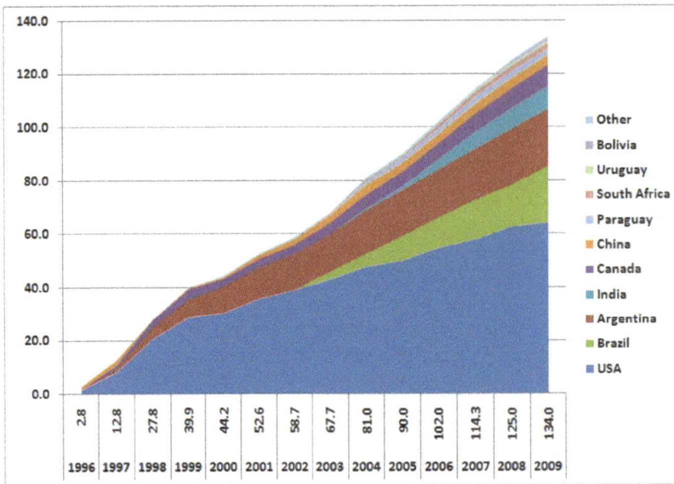

Land area used for genetically modified crops by country (1996–2009), in millions of hectares. In 2011, the land area used was 160 million hectares, or 1.6 million square kilometers.

In the US, by 2014, 94% of the planted area of soybeans, 96% of cotton and 93% of corn were genetically modified varieties. Genetically modified soybeans carried herbicide-tolerant traits only, but maize and cotton carried both herbicide tolerance and insect protection traits (the latter largely Bt protein). These constitute "input-traits" that are aimed to financially benefit the producers, but may have indirect environmental benefits and cost benefits to consumers. The Grocery Manufacturers of America estimated in 2003 that 70–75% of all processed foods in the U.S. contained a GM ingredient.

Europe grows relatively few genetically engineered crops with the exception of Spain, where one fifth of maize is genetically engineered, and smaller amounts in five other countries. The EU had a 'de facto' ban on the approval of new GM crops, from 1999 until 2004. GM crops are now regulated by the EU. In 2015, genetically engineered crops are banned in 38 countries worldwide, 19 of them in Europe. Developing countries grew 54 percent of genetically engineered crops in 2013.

In recent years GM crops expanded rapidly in developing countries. In 2013 approximately 18 million farmers grew 54% of worldwide GM crops in developing countries. 2013's largest increase was in Brazil (403,000 km^2 versus 368,000 km^2 in 2012). GM cotton began growing in India in 2002, reaching 110,000 km^2 in 2013.

According to the 2013 ISAAA brief: "...a total of 36 countries (35 + EU-28) have granted regulatory approvals for biotech crops for food and/or feed use and for environmental release or planting since 1994... a total of 2,833 regulatory approvals involving 27 GM crops and 336 GM events (NB: an "event" is a specific genetic modification in a specific species) have been issued by authorities, of which 1,321 are for food use (direct use or processing), 918 for feed use (direct use or processing) and 599 for environmental release or planting. Japan has the largest number (198), followed by the U.S.A. (165, not including "stacked" events), Canada (146), Mexico (131), South Korea (103), Australia (93), New Zealand (83), European Union (71 including approvals that have expired or under renewal process), Philippines (68), Taiwan (65), Colombia (59), China (55) and South Africa (52). Maize has the largest number (130 events in 27 countries), followed by cotton (49 events in 22 countries), potato (31 events in 10 countries), canola (30 events in 12 countries) and soybean (27 events in 26 countries).

Controversy

GM foods are controversial and the subject of protests, vandalism, referenda, legislation, court action and scientific disputes. The controversies involve consumers, biotechnology companies, governmental regulators, non-governmental organizations and scientists. The key areas are whether GM food should be labeled, the role of government regulators, the effect of GM crops on health and the environment, the effects of pesticide use and resistance, the impact on farmers, and their roles in feeding the world and energy production.

There is a scientific consensus that currently available food derived from GM crops poses no greater risk to human health than conventional food, but that each GM food needs to be tested on a case-by-case basis before introduction. Nonetheless, members of the public are much less likely than scientists to perceive GM foods as safe. The legal and regulatory status of GM foods varies by country, with some nations banning or restricting them, and others permitting them with widely differing degrees of regulation.

No reports of ill effects have been documented in the human population from GM food. Although GMO labeling is required in many countries, the United States Food and Drug Administration does not require labeling, nor does it recognize a distinction between approved GMO and non-GMO foods.

Advocacy groups such as Center for Food Safety, Union of Concerned Scientists, Greenpeace and the World Wildlife Fund claim that risks related to GM food have not been adequately examined and managed, that GMOs are not sufficiently tested and should be labelled, and that regulatory authorities and scientific bodies are too closely tied to industry. Some studies have claimed that genetically modified crops can cause harm; a 2016 review that reanalyzed the data from six of these studies found that their statistical methodologies were flawed and did not demonstrate harm, and said that conclusions about GMO crop safety should be drawn from "the totality of the evidence... instead of far-fetched evidence from single studies".

References

- Slater, Adrian; Scott, Nigel; Fowler, Mark (2008). Plant Biotechnology: the genetic manipulation of plants (2 ed.). Oxford, New York, USA: Oxford University Press Inc. ISBN 978-0-19-928261-6.
- Praitis, Vida (2006). "Creation of Transgenic Lines Using Microparticle Bombardment Methods". 351: 93–108. doi:10.1385/1-59745-151-7:93. ISBN 1-59745-151-7.
- Hayward, M.D.; Bosemark, N.O.; Romagosa, T. (2012). Plant Breeding: Principles and Prospects. Springer Science & Business Media. p. 131. ISBN 9789401115247.
- Martins VAP (2008). "Genomic Insights into Oil Biodegradation in Marine Systems". Microbial Biodegradation: Genomics and Molecular Biology. Caister Academic Press. ISBN 978-1-904455-17-2.
- James, Clive (2007). "Executive Summary". G lobal Status of Commercialized Biotech/GM Crops: 2007. ISAAA Briefs. 37. The International Service for the Acquisition of Agri-biotech Applications (ISAAA). ISBN 978-1-892456-42-7. OCLC 262649526. Archived from the original on 6 June 2008. Retrieved 24 September 2009.
- "A decade of EU-funded GMO research (2001–2010)" (PDF). Directorate-General for Research and Innovation. Biotechnologies, Agriculture, Food. European Commission, European Union. 2010. doi:10.2777/97784. ISBN 978-92-79-16344-9. Retrieved February 8, 2016.

- Panchin, Alexander Y.; Tuzhikov, Alexander I. (2016). "Published GMO studies find no evidence of harm when corrected for multiple comparisons". Critical Reviews in Biotechnology: 1–5. doi:10.3109/07388551.2015.1130684. PMID 26767435.
- Thomson JA. "Genetic Engineering of Plants" (PDF). Biotechnology. Encyclopedia of Life Support Systems. 3. Retrieved 17 July 2016.
- ISAAA 2015 Annual Report Executive Summary, 20th Anniversary (1996 to 2015) of the Global Commercialization of Biotech Crops and Biotech Crop Highlights in 2015 ISAAA Brief 51-2015, Retrieved 19 August 2016
- Final Report of the PABE research project (December 2001). "Public Perceptions of Agricultural Biotechnologies in Europe". Commission of European Communities. Retrieved February 24, 2016.
- Bashshur, Ramona (February 2013). "FDA and Regulation of GMOs". American Bar Association. Retrieved February 24, 2016.
- Lynch, Diahanna; Vogel, David (April 5, 2001). "The Regulation of GMOs in Europe and the United States: A Case-Study of Contemporary European Regulatory Politics". Council on Foreign Relations. Retrieved February 24, 2016.

5

Applications of Horticulture in Different Fields

This chapter discusses the various applications of horticulture in fields like horticultural botany, container gardening, garden designing, urban horticulture etc. These fields use the advancements, technology and techniques of horticulture for various purposes. The content included in this chapter gives the reader in-depth knowledge about the practical applications of horticulture.

Horticultural Botany

Horticultural botany is the study of the botany of current and potential cultivated plants, with emphasis on the ornamental plants of horticulture, by a horticultural botanist or *plantsman*—plantsperson."

Horticultural Botanists

Professional horticultural botanists are employed by botanical gardens, large plant nurseries, university departments, and government agencies. The activities vary according to the mission goals and priorities of the institutions.

Duties can include:

- searching for new plants suitable for cultivation (*plant hunting*),
- communicating with and advising the general public on matters concerning the classification and botanical nomenclature of cultivated plants,
- carrying out original research on these topics:
    - describing the cultivated plants and their history of particular regions in horticultural floras,
    - recording new plant introductions;
- maintaining databases of cultivated plants;
- curating horticultural herbaria, including collections of dried specimens and images;
- contributing to the *International Code of Nomenclature for Cultivated Plants*.

Taxonomy of Cultivated Plants

Perhaps the single major concern of horticultural botanists has been the taxonomy of cultivated plants and so, at the present time, many horticultural botanists will be members if the *International Association of Cultivated Plant Taxonomy* which was established in 2007.

Horticultural Flora

A horticultural flora, also known as a garden flora, is a plant identification aid structured in the same way as a native plants flora. It serves the same purpose: to facilitate plant identification; however, it only includes plants that are under cultivation as ornamental plants growing within the prescribed climate zone or region. Traditionally published in book form, often in several volumes, such floras are increasingly likely to be produced as websites or CD ROMs.

Scope and Contents

Horticultural floras include both cultigens (plants deliberately altered in some way by human activity) and those wild plants brought directly into cultivation that do not have cultigen names. They might also include colour images and useful information specific to the zone or region including:

- historical details about outstanding public and private cultivated plant collections
- exceptional trees (age, history, rarity, size etc.)
- prominent nurserymen and plant breeders
- references to the taxonomic and other literature on the plant groups
- easy "spotting" or "field" characters useful for quick identification
- notes on ecology (especially the potential of plants to naturalise and become invasive)
- horticultural history of introduction
- conservation.

Uses

Written by a professional plant taxonomist or plantsperson, a horticultural flora assists clarification of scientific and common names, the identification of plant characteristics that occur in cultivated plants that are additional to those in wild counterparts, and descriptions of cultivars.

Although horticultural floras may include a range of food plants, their emphasis is generally on ornamental plants and so these floras are sometimes referred to as "garden floras". Increasingly they provide data for sustainable landscaping, such as:

- drought tolerance and irrigation needs
- food sources of native and migratory bird and butterfly species
- companion planting
- invasive species notation
- local habitat restoration aspects.

Published

Numerous encyclopaedic listings of cultivated plants have been compiled but only four substantial horticultural floras have ever been produced, these being for: North America; Europe; South-eastern Australia, Hawaii and the tropics.

There are several publications on trees which follow the format of botanical keys and descriptions for the trees of a specific region, notably for North America and California.

Garden Design

The White Garden at Sissinghurst Castle Garden, designed by Vita Sackville-West

Garden design is the art and process of designing and creating plans for layout and planting of gardens and landscapes. Garden design may be done by the garden owner themselves, or by professionals of varying levels of experience and expertise. Most professional garden designers have some training in horticulture and the principles of design, and some are also landscape architects, a more formal level of training that usually requires an advanced degree and often a state license. Amateur gardeners may also attain a high level of experience from extensive hours working in their own gardens, through casual study, serious study in Master Gardener Programs, or by joining gardening clubs.

Elements of Garden Design

Whether a garden is designed by a professional or an amateur, certain principles form the basis of effective garden design, resulting in the creation of gardens to meet the needs, goals and desires of the users or owners of the gardens.

Elements of garden design include the layout of hard landscape, such as paths, walls, water features, sitting areas and decking; as well as the plants themselves, with consideration for their horticultural requirements, their season-to-season appearance, lifespan, growth habit, size, speed of growth, and combinations with other plants and landscape features. Consideration is also given to the maintenance needs of the garden, including the time or funds available for regular maintenance, which can affect the choice of plants in terms of speed of growth, spreading or self-seeding of the plants, whether annual or perennial, and bloom-time, and many other characteristics.

Important considerations in garden design include how the garden will be used, the desired stylistic genre (formal or informal, modern or traditional etc.), and the way the garden space will connect to the home or other structures in the surrounding areas. All of these considerations are subject to the limitations of the prescribed budget.

Location

A garden's location can have a substantial influence on its design. Topographical landscape features such as steep slopes, vistas, hills and outcrops etc. may suggest or determine aspects of design such as layout, and can be used and augmented in order to create a particular impression. The soils of the site will affect what types of plant may be grown, as will the garden's climate zone and various microclimates. The locational context of the garden can also influence its design; for example an urban setting may require a different design style to a rural one. Similarly, a windy coastal location may necessitate a different treatment compared to a sheltered inland site.

Soil

The quality of a garden's soil can have a significant influence on a garden's design and its subsequent success. Soil influences the availability of water and nutrients, the activity of soil micro-organisms, and temperature within the root zone, and thus may have a determining effect on the types of plants which will grow successfully in the garden. However soils may be replaced or improved in order to make them more suitable.

Traditionally, garden soil is improved by amendment, the process of adding beneficial materials to the native subsoil and particularly the topsoil. The added materials, which may consist of compost, peat, sand, mineral dust, or manure, among others, are mixed with the soil to the preferred depth. The amount and type of amendment may depend on many factors, including the amount of existing soil humus, the soil structure (clay, silt, sand, loam etc.), the soil acidity/alkalinity, and the choice of plants to be grown. One source states that, "conditioning the soil thoroughly before planting enables the plants to establish themselves quickly and so play their part in the design." However, not all gardens are, or should be, amended in this manner, since many plants prefer an impoverished soil. In this case, poor soil is better than a rich soil that has been artificially enriched.

Boundaries

The design of a garden can be affected by the nature of its boundaries, both external and internal, and in turn the design can influence the boundaries, including via creation of new ones. Planting can be used to modify an existing boundary line by softening or widening it. Introducing internal boundaries can help divide or break up a garden into smaller areas.

The main types of boundary within a garden are hedges, walls and fences. A hedge may be evergreen or deciduous, formal or informal, short or tall, depending on the style of the garden and purpose of the boundary. A wall has a strong foundation beneath it at all points, and is usually - but not always - built from brick, stone or concrete blocks. A fence differs from a wall in that it is anchored only at intervals, and is usually constructed using wood or metal (such as iron or wire mesh).

Boundaries may be constructed for several reasons: to keep out livestock or intruders, to provide privacy, to create shelter from strong winds and provide microclimates, to screen unattractive structures or views, and to create an element of surprise.

Surfaces

In temperate western gardens, a smooth expanse of lawn is often considered essential to a garden. However garden designers may use other surfaces, for example those “made up of loose gravel, small pebbles, or wood chips” in order to create a different appearance and feel. Designers may also utilise the contrast in texture and color between different surfaces in order to create an overall pattern in the design.

Naturalistic planting design

Surfaces for paths and access points are chosen for practical as well as aesthetic reasons. Issues such as safety, maintenance and durability may need to be considered by the designer. Gardens designed for public access need to cope with heavier foot traffic and hence may utilise surfaces - such as resin-bonded gravel - that are rarely used in private gardens.

Planting Design

Planting design requires design talent and aesthetic judgement combined with a good level of horticultural, ecological and cultural knowledge. It includes two major traditions: formal rectilinear planting design (Persia and Europe); and formal asymmetrical (Asia) and naturalistic planting design.

History

Persian gardens are credited with originating aesthetic and diverse planting design. A correct Persian garden will be divided into four sectors with water being very important for both irrigation and aesthetics. The four sectors symbolize the Zoroastrian elements of sky, earth, water and plants. Planting in ancient and Medieval European gardens was often a mix of herbs for medicinal use, vegetables for consumption, and flowers for decoration. Purely aesthetic planting layouts developed after the Medieval period in Renaissance gardens, as are shown in late-renaissance paintings and plans. The designs of the Italian Renaissance garden were geometrical and plants were used to form spaces and patterns. The gardens of the French Renaissance and Baroque Garden à la française era continued the 'formal garden' planting aesthetic.

In Asia the asymmetrical traditions of planting design in Chinese gardens and Japanese gardens originated in the Jin Dynasty (265–420) of China. The gardens' plantings have a controlled but naturalistic aesthetic. In Europe the arrangement of plants in informal groups developed as part of the English Landscape Garden style, and subsequently the French landscape garden, and was strongly influenced by the picturesque art movement.

Application

A planting plan gives specific instructions, often for a contractor about how the soil is to be prepared, what species are to be planted, what size and spacing is to be used and what maintenance operations are to be carried out under the contract. Owners of private gardens may also use planting plans, not for contractual purposes, as an aid to thinking about a design and as a record of what has been planted. A planting strategy is a long term strategy for the design, establishment and management of different types of vegetation in a landscape or garden.

Garden chairs and table

Planting can be established by directly employed gardeners and horticulturalists or it can be established by a landscape contractor (also known as a landscape gardener). Landscape contractors work to drawings and specifications prepared by garden designers or landscape architects.

Garden Furniture

Garden furniture may range from a *patio set* consisting of a table, four or six chairs and a parasol, through benches, swings, various lighting, to stunning artifacts in brutal concrete or weathered oak. Patio heaters, that run on bottled butane or propane, are often used to enable people to sit outside at night or in cold weather. A picnic table, is used for the purpose of eating a meal outdoors such as in a garden.

The materials used to manufacture modern patio furniture include stones, metals, vinyl, plastics, resins, glass, and treated woods.

Sunlight

While sunlight is not always easily control by the gardener, it is an important element of garden design. The amount of available light is a critical factor in determining what plants may be grown. Sunlight will, therefore, have a substantial influence on the character of the garden. For example, a rose garden is generally not successful in full shade, while a garden of hostas may not thrive in hot sun.

As another example, a vegetable garden may need to be placed in a sunny location, and if that location is not ideal for the overall garden design goals, the designer may need to change other aspects of the garden.

In some cases, the amount of available sunlight can be influenced by the gardener. The location of trees, other shade plants, garden structures, or, when designing an entire property, even buildings, might be selected or changed based on their influence in increasing or reducing the amount of sunlight provided to various areas of the property.

In other cases, the amount of sunlight is not under the gardener's control. Nearby buildings, plants on other properties, or simply the climate of the local area, may limit the available sunlight. Or, substantial changes in the light conditions of the garden may not be within the gardener's means. In this case, it is important to plan a garden that is compatible with the existing light conditions.

Lighting

Garden lighting can be an important aspect of garden design. In most cases, various types of lighting techniques may be classified and defined by heights: safety lighting, uplighting, and downlighting. Safety lighting is the most practical application. However, it is more important to determine the type of lamps and fittings needed to create the desired effects. Light regulates three major plant processes: photosynthesis, phototropism, and photoperiodism.

Photosynthesis provides the energy required to produce the energy source of plants.

Phototropism is the effect of light on plant growth that causes the plant to grow toward or away from the light. Photoperiodism is a plant's response or capacity to respond to photoperiod, a recurring cycle of light and dark periods of constant length.

Types of Gardens

Moorish Generalife courtyard fountain

Islamic Gardens

Garden design, and the Islamic garden tradition, began with creating the Paradise garden in Ancient Persia, in Western Asia. It evolved over the centuries, and in the different cultures Islamic dynasties came to rule in Central— South Asia, the Near East, North Africa, and the Iberian Peninsula.

Examples

Some styles and examples include:

- Persian gardens
    - Eram Garden
    - Fin Garden
- Mughal gardens
    - Nishat Bagh
    - Shalimar Gardens (Lahore)
    - Yadavindra Gardens (Pinjore)
- Charbagh
    - Taj Mahal
    - Tomb of Humayun gardens
- Bagh (garden)
    - Bagh-e Babur
    - Shalimar Bagh (Srinagar)
- Al-Andalus—Moorish architecture and gardens
    - Alcázar of Seville

- Alhambra
- Generalife

Andalusian Patio of Córdoba, Spain

Mediterranean Gardens

Garden design history and precedents from the Mediterranean region include:

- Ancient Greek and Hellenistic gardens
- Ancient Roman gardens
    - Peristyle gardens —*evolved into Cloister gardens.*
    - House of the Vettii — *Pompeii.*
    - Horti Sallustiani
- Byzantine gardens
- Spanish gardens
    - Andalusian Patio

Renaissance and Formal Gardens

French formal garden and parterre at Château de Villandry in the Loire Valley

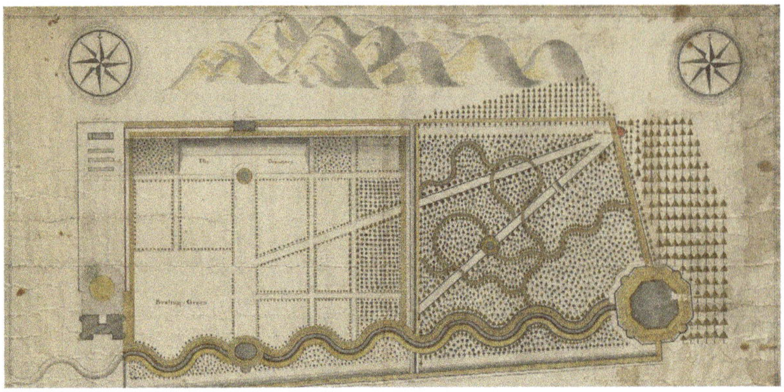

A plan of a formal garden for a country estate in Wales, 1765

A formal garden in the Persian garden and European garden design traditions is rectilinear and axial in design. The equally formal garden, without axial symmetry (asymmetrical) or other geometries, is the garden design tradition of Chinese gardens and Japanese gardens. The Zen garden of rocks, moss and raked gravel is an example. The Western model is an ordered garden laid out in carefully planned geometric and often symmetrical lines. Lawns and hedges in a formal garden need to be kept neatly clipped for maximum effect. Trees, shrubs, subshrubs and other foliage are carefully arranged, shaped and continually maintained. A French garden or Garden à la française, is a specific kind of formal garden, laid out in the manner of André Le Nôtre; it is centered on the façade of a building, with radiating avenues and paths of gravel, lawns, parterres and pools (*bassins*) of reflective water enclosed in geometric shapes by stone coping, with fountains and sculpture.

The Garden à la française style has origins in fifteenth-century Italian Renaissance gardens, such as the Villa d'Este, Boboli Gardens, and Villa Lante in Italy. The style was brought to France and expressed in the gardens of the French Renaissance. Some of the earliest formal parterres of clipped evergreens were those laid out at Anet by Claude Mollet, the founder of a dynasty of nurserymen-designers that lasted deep into the 18th century. The Gardens of Versailles are an ultimate example of Garden à la française, composed of many different distinct gardens, and designed by André Le Nôtre.

English Renaissance gardens in a rectilinear formal design were a feature of the stately homes. The introduction of the parterre was at Wilton House in the 1630s. In the early eighteenth century, the publication of Dezallier d'Argenville, *La théorie et la pratique du jardinage* (1709) was translated into English and German, and was the central document for the later formal gardens of Continental Europe.

Traditional formal Spanish garden design evolved with Persian garden and European Renaissance garden influences. The internationally renowned Alhambra and Generalife in Granada, built in the Moorish Al-Andalus era, have influenced design for centuries. The Ibero-American Exposition of 1929 World's Fair in Seville, Spain was located in the celebrated Maria Luisa Park (*Parque de Maria Luisa*) designed by Jean-Claude Nicolas Forestier.

Formal gardening in the Italian and French manners was reintroduced at the turn of the twentieth century. Beatrix Farrand's formal Italian garden areas at Dumbarton Oaks in Washington, D.C., and Achille Duchêne's restored French water parterre at Blenheim Palace in England are examples of the modern formal garden. The Conservatory Garden in Central Park of New York City features a formal garden, as do many other parks and estates such as Filoli in California.

The simplest formal garden would be a box-trimmed hedge lining or enclosing a carefully laid out flowerbed or garden bed of simple geometric shape, such as a knot garden. The more developed and elaborate formal gardens contain statuary and fountains.

Formal garden laid out at the Abbaye de Valloires, Picardy, by Gilles Clément, 1987

Features in a formal garden may include:

- Terrace
- Topiary
- Statuary
- Hedge
- Bosquet
- Parterre
- Sylvan theater
- Pergola
- Pavilion
- Landscaping

English Landscape and Naturalistic Gardens

The English Landscape Garden style practically swept away the geometries of earlier English and European Renaissance formal gardens. William Kent and Lancelot "Capability" Brown were leading proponents, among many other designers. The naturalistic English Garden style (French: *Jardin anglais*, Italian: *Giardino all'inglese*, German: *Englischer Landschaftsgarten*) of the 1730s and on transformed private and civic garden design across Europe. The French Landscape Garden subsequently continued the style's development on the Continent.

Cottage Gardens

A cottage garden uses an informal design, traditional materials, dense plantings, and a mixture of ornamental and edible plants. Cottage gardens go back many centuries, but their popularity grew in 1870s England in response to the more structured Victorian English estate gardens that used restrained designs with massed beds of brilliantly colored greenhouse annuals. They are more casual by design, depending on grace and charm rather than grandeur and formal structure. The influential British garden authors and designers, William Robinson at Gravetye Manor in Sussex, and Gertrude Jekyll at Munstead Wood in Surrey, both wrote and gardened in England. Jekyll's series of thematic gardening books emphasized the importance and value of natural plantings were an influence in Europe and the United States. Also influential half a century later was Margery Fish, whose surviving garden at East Lambrook Manor emphasizes, among other things, native plant life and the natural patterns produced by self-spreading and self-seeding.

The earliest cottage gardens were far more practical than modern versions—with an emphasis on vegetables and herbs, along with fruit trees, beehives, and even livestock if land allowed. Flowers were used to fill any spaces in between. Over time, flowers became more dominant. Modern day cottage gardens include countless regional and personal variations of the more traditional English cottage garden.

Kitchen Garden or Potager

The traditional kitchen garden, also known as a potager, is a seasonally used space separate from the rest of the residential garden - the ornamental plants and lawn areas. Most vegetable gardens are still miniature versions of old family farm plots with square or rectangular beds, but the kitchen garden is different not only in its history, but also its design.

The kitchen garden may be a landscape feature that can be the central feature of an ornamental, all-season landscape, but can be little more than a humble vegetable plot. It is a source of herbs, vegetables, fruits, and flowers, but it is also a structured garden space, a design based on repetitive geometric patterns.

Formal potager at Villandry, France

An illustration from Walter Crane's 1906 book,*Flowers from Shakespeare's Garden: a Posy from the Plays*

The kitchen garden has year-round visual appeal and can incorporate permanent perennials or woody plantings around (or among) the annual plants.

Shakespeare Garden

A Shakespeare garden is a themed garden that cultivates plants mentioned in the works of William Shakespeare. In English-speaking countries, particularly the United States, these are often public gardens associated with parks, universities, and Shakespeare festivals. Shakespeare gardens are sites of cultural, educational, and romantic interest and can be locations for outdoor weddings.

Signs near the plants usually provide relevant quotations. A Shakespeare garden usually includes several dozen species, either in herbaceous profusion or in a geometric layout with boxwood dividers. Typical amenities are walkways and benches and a weather-resistant bust of Shakespeare. Shakespeare gardens may accompany reproductions of Elizabethan architecture. Some Shakespeare gardens also grow species typical of the Elizabethan period but not mentioned in Shakespeare's plays or poetry.

Rock Garden

A rock garden, also known as a rockery or an alpine garden, is a type of garden that features extensive use of rocks or stones, along with plants native to rocky or alpine environments.

A naturalistic rockery in England

Rock garden in Chandigarh, India

Rock garden plants tend to be small, both because many of the species are naturally small, and so as not to cover up the rocks. They may be grown in troughs (containers), or in the ground. The plants will usually be types that prefer well-drained soil and less water.

The usual form of a rock garden is a pile of rocks, large and small, aesthetically arranged, and with small gaps between, where the plants will be rooted. Some rock gardens incorporate bonsai.

Some rock gardens are designed and built to look like natural outcrops of bedrock. Stones are aligned to suggest a bedding plane and plants are often used to conceal the joints between the stones. This type of rock garden was popular in Victorian times, often designed and built by professional landscape architects. The same approach is sometimes used in modern campus or commercial landscaping, but can also be applied in smaller private gardens.

The Japanese rock garden, in the west often referred to as *Zen garden*, is a special kind of rock garden which contains few plants. Rock gardens have become increasingly popular as landscape features in tropical countries such as Thailand. The combination of wet weather and heavy shade trees, along with the use of heavy plastic liners to stop unwanted plant growth, has made this type of arrangement ideal for both residential and commercial gardens due to its easier maintenance and drainage.

East Asian Gardens

Japanese and Korean gardens, originally influenced by Chinese gardens, can be found at Buddhist temples and historic sites, private homes, in neighborhood or city parks, and at historical landmarks such as Buddhist temples. Some of the Japanese gardens most famous in the Western world and Japan are gardens in the *karesansui* (rock garden) tradition. The Ryōan-ji temple garden is a well-known example. There are Japanese gardens of various styles, with plantings and often evoking *wabi sabi* simplicity. In Japanese culture, garden-making is a high art, intimately linked to the arts of calligraphy and ink painting.

Contemporary Garden

Contemporary garden

Contemporary water feature

The contemporary style garden has gained popularity in the UK in the last 10 years. This is partly due to the increase of modern housing with small gardens as well as the cultural shift towards contemporary design. This style of garden can be defined by the use 'clean' design lines, with focus on hard landscaping materials like stone, hardwood, rendered walls. Planting style is bold but simple with the use of drifts of one or two plants that repeat throughout the design. Grasses are a very popular choice for this style of design. Lighting effects also play an integral role in the modern garden. Subtle lighting effects can be achieved with the use of carefully placed low voltage LED lights incorporated into paving and walls.

Residential Gardens

A residential or private domestic garden, is the most common form of garden and is in proximity to a residence, such as the 'front garden' or 'back garden'. The front garden may be a formal and semi-public space and so subject to the constraints of convention and local laws. While typically found in the yard of the residence, a garden may also be established on a roof, in an atrium or courtyard, on a balcony, in windowboxes, or on a patio. Residential gardens are typically designed at human scale, as they are most often intended for private use. However, the garden of a great house or a large estate may be larger than a public park.

Residential gardens may feature specialized gardens, such as those for exhibiting one particular type of plant, or special features, such as rockery or water features. They are also used for growing herbs and vegetables and are thus an important element of sustainability.

Container Garden

Container gardening or pot gardening is the practice of growing plants, including edible plants, exclusively in containers instead of planting them in the ground.

Container garden on front porch

A container is the general term used in gardening for a small, enclosed and usually portable object used for displaying live flowers or plants. It may take the form of a pot, box, tub, pot, basket, tin, barrel or hanging basket.

Pots, traditionally made of terracotta but now more commonly plastic, and window boxes have been the most commonly seen. Small pots are commonly called flowerpots. In some cases, this method of growing is used for ornamental purposes. This method is also useful in areas where the soil or climate is unsuitable for the plant or crop in question. Using a container is also generally necessary for houseplants. Limited growing space, or growing space that is paved over, can also make this option appealing to the gardener.

Planting

Containers range from simple plastic pots, teacups to complex automatic-watering irrigation systems. This flexibility in design is another reason container gardening is popular with growers. They can be found on porches, front steps, and in urban locations, on rooftops. Sub-irrigated planters (SIP) are a type of container that may be used in container gardens. Potting material must be loose and allow drainage to offer proper aeration for roots to breathe, preventing root rot.

Re-potting

Re-potting is the action of placing an already potted plant into a larger or smaller pot. A pot that fits a plant's root system better is normally used. Plants are usually re-potted according to the size of their root system. Most plants need to be re-potted every few years, because they become "pot-bound" or "root-bound." Plants' roots can sense its surroundings, including the size of the pot it is in, and increasing the pot size allows plant size to increase proportionally.

Root-bound

Advantages

Many types of plants are suitable for the container, including decorative flowers, herbs, cacti, vegetables, and small trees. There are many advantages to growing plants in containers, namely:

- Less risk of soil-borne disease
- Virtually eliminate weed problems
- Mobile plants gives more control over moisture, sunlight & temperature

Pot Farming

Tomato plants growing in a pot farming alongside a small house in New Jersey in fifteen garbage cans filled with soil, grew over 700 tomatoes during the summer of 2013.

Crops can be grown in flowerpots and it is called pot farming, pot agriculture or pot horticulture.

Houseplant

A houseplant is a plant that is grown indoors in places such as residences and offices. Houseplants are commonly grown for decorative purposes, positive psychological effects, keeping fresh or health reasons such as indoor air purification. Plants used in this fashion are most commonly, though not always, tropical or semi-tropical epiphytes, succulents or cacti.

Houseplants need the correct moisture, light levels, soil mixture, temperature, and humidity. As well, houseplants need the proper fertilizer and correct-sized pots.

A variegated cultivar of *Sansevieria trifasciata* (namely 'Laurentii'), a common houseplant

Plant Requirements

Major factors that should be considered when caring for houseplants are moisture, light, soil mixture, temperature, humidity, fertilizers, potting, and pest control. The following includes some general guidelines for houseplant care. For specific houseplant needs, the tags that sometimes come with plants are notoriously unhelpful and generic. Specific care information may be found widely online and in books.

Moisture

Succulents, or water-retaining plants, such as this jelly bean plant (*Sedum rubrotinctum*), are often grown as houseplants

Both under-watering and over-watering can be detrimental to a houseplant. The best way to determine whether a plant needs water is to check the soil moisture. Feeling the soil is most reliable, since moisture meters are often inaccurate. Most potted plants must be allowed to reach an appropriate level of dryness in between waterings, though the amount of watering required varies greatly depending on the species. Proper soil moisture can range from still slightly moist on the soil surface to very dry to nearly the bottom of the pot. Watering a plant by the calendar is not recommended. If a plant does need to be watered, water should be slowly poured over the surface of the soil until it begins to drain out the bottom of the pot, ensuring complete saturation. However, sometimes the soil separates from the sides of the pot if it is allowed to dry out thoroughly, allowing the water to flow down the sides of the rootball and out the bottom too quickly to be absorbed and retained by the soil and roots. If this is the case, it may be necessary to set the plant in a shallow dish of water long enough for it to soak up enough water to moisten the rootball to its center. Repotting should eliminate this problem. Repotting should be done only when necessary, since the roots of a plant that is in an excessively large pot may rot.

Light

A skylight provides sun to these plants in the dining hall of Currier House at Harvard University

Through the process of photosynthesis, plants convert the energy in sunlight to chemical energy, which fuels plant growth. The two important factors for providing light to a house plant are *intensity* and *duration*.

Different plants require different light intensities. Intensity (or quality) of light is difficult to measure without a light meter. It is usually measured in units of lux. 100 lux or less is usually considered "low intensity" or "indirect" lighting. A bright office has about 400 lux of illumination. 1,000 lux or more is usually considered "high intensity" lighting. Direct outdoor sunlight is in the range 32,000-100,000 lux. Foot-candles are also occasionally used.

The duration of light exposure is as important as the intensity. Quality exposure of 8 to 16 hours is ideal for most plants. Photoperiodism must also be considered, since some plants such as Poinsettia and Schlumbergera are influenced by either decreasing or increasing daylight hours.

Some potted plants can grow to very large sizes, such as this tree found in Auckland Airport.

Windows are the most common sources of light for houseplants. In the Northern Hemisphere, south-facing windows have the most sun exposure, while western, eastern, and north-facing windows have progressively less exposure. Natural sunlight through windows is affected by seasonal changes, cloud cover, and window treatments. The length of time that light is provided will determine how the plant grows. Providing 16 hours of light/day will promote strong roots, stems and abundant leaves. Decreasing that amount to 12 hours of light/day will signal that the short days of Winter are coming so the plant energy will focus more on flower production and less on green growth.

Artificial light sources can provide an alternative or supplement to window lighting. Fluorescent lighting provides excellent light quality whereas standard incandescent bulbs do little to promote plant growth. "Cool", or "blue", fluorescent lights at 6500k provide the light needed for lush green foliage plants, while "warm", or "red", fluorescent lights at 3000k provide the light needed for blooming flowers and fruit production. Warm whites are better for flowering plants while cool whites are more suitable for green, leafy growth. When used together, these bulbs are closer to the full spectrum light that comes from the sun, although less powerful.

There are several types of lighting units which can sustain indoor plants, Fluorescent, Halide or Diode. Choosing the best type depends on the need of the plants grown and/or your budget. For the sake of efficiency, incandescents, no matter how cheap, should not be used. Not only do they provide little or no benefit to plants, but the cost of the electricity will outweigh the cost of purchasing multiple incandescent bulbs and fixtures in the long run. Fluorescents are mass marketed. CFL fluorescents are the cheapest option, but more than a couple bulbs are almost always required to be running at once. The next step up is shop lights, available mostly in 2-ft (609 mm) or 4-ft (1219 mm) fluorescent tubes. The best type of fluorescent are called High Output Fluorescents. Also available in 2-ft (609 mm) and 4-ft(1219 mm) tubes, these bulbs provide more wattage (54 watts each is standard) thus more lumens per watt (92 at the current time). That is about 5,000 lumens per dual 4-ft (1219 mm) H.O. Fluorescent Unit. The most serious lights that are used by professionals and in greenhouses as a supplement, are known as Metal Halide or High Pressure Sodium Grow Lights. These lights provide the most lumens, heat and intensity of light so they should be positioned respectively further from the tops of plants to prevent burning. LED or Light Emitting Diode grow lights which produce the photosynthetic optimum red (640-670 nm) and blue (430-460 nm), they have long life expectancies, and efficiency, 100 watts producing 3,400 lumens output from a unit 400x212x62mm, they are also rather cool, so you can have them quite close to the plants.

Soil

Houseplants are generally grown in specialized soils called *potting compost* or *potting soil*, not in local natural soil. A good potting compost mixture includes soil conditioners to provide the plant with nutrients, support, adequate drainage, and proper aeration. Most potting composts contain a combination of peat and vermiculite or perlite. Concern over environmental damage to peat bogs, however, is leading to the replacement of peat by coir (coconut fibre), which is a sustainable resource. A nutrient rich compost can usually be bought wherever potted plants are sold.

If local natural soil is to be used, it should first be heat sterilized by placing the soil in an oven at 90 °C (200 °F) for at least 30 minutes. This will ensure that the soil does not contain any harmful bacteria. Most locals soils, especially those with a high proportion of clay, do not drain well enough to be a suitable growing medium for houseplants. Coir or peat is used to increase aeration and make heavy soils more absorbent. Vermiculite and perlite aid in drainage in a soil mixture. Perlite is recommended over vermiculite because it does not break down as easily. A coarse grade sand or grit can be used as a substitute for a drainage mechanism if needed. These three ingredients can be mixed in varying ratios to create different potting soil types. For a plant that requires fast drainage, such as a cactus, use plenty of coarse sand, grit or perlite. For a plant that requires plenty of moisture, use more coir. A good all purpose soil mixture is 2 parts coir and 1 part perlite or vermiculite. A so-called "heavy soil mix" will contain sterilised soil, milled sphagnum moss or coir, and perlite (or vermiculite) in equal proportions. It is also possible to make a soil mixture that actually contains no soil by mixing equal parts peat moss and perlite (or vermiculite). The soiless mixture will retain more moisture.

Temperature

Tillandsias. Air plants survive in 50-90 degrees. Softened water (aka most tap water) will kill them because of the salt.

Most houseplants are tropical species selected for their adaptation to growth in a climate which ranges from 15 °C to 25 °C (60 °F to 80 °F), similar to the temperature in most homes. Temperature control for other plants with differing requirements needs attention to heating and/or cooling.

Echeveria are a houseplant commonly found in the wild of South Africa.

Succulents are a houseplant commonly used as wedding favors.

Humidity

Humidity is slightly more difficult to control than temperature. The more commonly used houseplants have established that they can survive in low humidity environments as long as their roots are kept properly irrigated. Most plants thrive in 80% relative humidity while most homes are usually kept around 20% to 60% relative humidity. Besides buying a humidifier, there are a few things that can be done to increase humidity around houseplants. The most popular methods used to raise the ambient humidity are misting and pebble trays, which are shallow trays covered with pebbles and filled with water that evaporates to increase humidity. Other methods of raising humidity include grouping plants closely together and not placing plants in drafty areas. Misting has become controversial among gardeners and horticulturists, some of whom will point out that it seldom provides moisture in the form of water vapor, and that, instead, by wetting the leaves it increases disease problems.

Fertilizers

In a potted environment, soil nutrients can eventually deplete. Adding fertilizer can artificially provide these nutrients. However, adding unnecessary fertilizer can be harmful to the plant. Because of this, careful consideration must be taken before fertilizing. If a plant has been in the same potting mix for a year or more and is no longer thriving, then it may be a candidate for nutrient replacement done by using a complete fertilizer at half the recommended label dilution rate.

Fertilizers are usually marked with a number such as 20–20–20. These numbers indicate the percentages of nitrogen, phosphorus, and potassium respectively, the three elements that are needed in the most quantity for plant growth. Nitrogen is essential for green, leafy growth. Phosphorus is essential for flowering or fruiting plants. Potassium is essential for strong roots and increased nutrient uptake. Numbers higher than 15 are usually man-made, chemical fertilizers. Organic fertilizers have a much lower ratio. A 4–2–2 ratio of these elements is usually good for green foliage plants, while a 2–6–4 ratio is usually better for flowering plants. A complete fertilizer will also include the minor and trace elements, such as calcium, magnesium and iron.

While variation may occur between brands, a general rule is to mix 1 tablespoon to every four liters (one imperial gallon) of water. In all cases, it is better to under-fertilize than over-fertilize. The diluted mixture is then used to water the plants. The growth of the plants should be monitored to determine if the fertilizer is helping or harming, and how often (if at all) it should be used. Schedules can range from every other week to every three months. For convenience, granular, time-released fertilizers are also available.

Crassula ovata in a clay container (Italian terra cotta).

Pot Types and Sizes

Proper pot size is an important factor to consider. A pot that is too large will cause root disease because of the excess moisture retained in the soil, while a pot that is too small will restrict a plant's growth. Generally, a plant can stay in the same pot for two or so years. Pots come in a variety of types as well, but usually can be broken down into two groups: porous and non-porous. Porous pots are usually clay and are highly recommended because they provide better aeration as air passes laterally through the sides of the pot. Non-porous pots such as glazed or plastic pots tend to hold moisture longer and restrict airflow. Another needed feature is drainage holes. Usually pots come with holes in the bottom to allow excess water to flow out of the soil which helps to prevent root rot.

If a pot does not have drainage holes, it is best to double pot that plant so the inner pot can be lifted out and the excess water accumulated in the bottom of the outer pot can be removed. Soak old pots thoroughly in a solution of 1 part bleach to 10 parts water to kill any bacteria that may remain.

Effect on Indoor Air Pollution

Indoor plants reduce components of indoor air pollution, particularly volatile organic compounds (VOC) such as benzene, toluene, and xylene. The compounds are removed primarily by soil microorganisms. Plants can also remove CO_2 "Carbon dioxide", which is correlated with lower work performance, from indoor areas. The effect has been investigated by NASA for use in spacecraft. Plants also appear to reduce airborne microbes and increase humidity.

Alternative Growing Methods

Hydroponics

Aside from traditional soil mixtures, media such as expanded clay may be employed in hydroponics, in which the plant is grown in a water and nutrient solution. This technique has a number of benefits, including an odorless, reusable, and more hygienic media. Any habitat for soil-bound pests is also eliminated, and the plant's water supply is less variable. However, some plants do not grow well with this technique, and media is often difficult to find in some parts of the world, such as North America, where hydroponics and specifically hydroculture is not as well-known or widespread.

Subirrigation

Subirrigation offers another alternative to top-watering techniques. In this approach the plant is watered from the bottom of the pot. Water is transferred up into the potting media (be it soil or others) by capillary action. Advantages of this technique include controlled amounts of water, resulting in lower chances of overwatering if done correctly, no need to drain plants after watering unlike traditional top-water methods, and less compaction of the media due to the pressure put on the media from top-watering.

List of Common Examples

Cyclamen sp.

Dracaena braunii

Tropical and Subtropical

- *Aglaonema* (Chinese Evergreen)
- *Alocasia*
- *Amaryllis*
- *Aphelandra squarrosa* (Zebra Plant)
- *Araucaria heterophylla* (Norfolk Island Pine)
- *Asparagus aethiopicus* (Asparagus Fern)
- *Begonia* species and cultivars
- *Bromeliaceae* (Bromeliads)
- *Chamaedorea elegans* (Parlor Palm)
- *Chlorophytum comosum* (Spider Plant)
- *Citrus*, compact cultivars such as the Meyer Lemon
- *Dracaena*
- *Dieffenbachia* (Dumbcane)
- *Epipremnum aureum* (Golden Pothos)
- *Ficus benjamina* (Weeping Fig)
- *Ficus elastica* (Rubber Plant)
- *Hippeastrum*
- *Mimosa pudica* (Sensitive Plant)

- *Nephrolepis exaltata cv. Bostoniensis* (Boston Fern)
- *Orchidaceae* (The orchids)
    - *Cattleya* and intergeneric hybrids thereof (e.g. Brassolaeliocattleya, Sophrolaelio-cattleya)
    - *Cymbidium*
    - *Dendrobium*
    - *Miltoniopsis*
    - *Oncidium*
    - *Paphiopedilum*
    - *Phalaenopsis*
- *Peperomia* species
- *Philodendron* species
- *Maranta* (The Prayer Plants)
- *Saintpaulia* (African violet)
- *Sansevieria trifasciata* (Mother-inlaw's tongue)
- *Schefflera arboricola* (Umbrella Plant)
- *Sinningia speciosa* (Gloxinia)
- *Spathiphyllum* (Peace Lily)
- *Stephanotis floribunda* (Madagascar Jasmine)
- *Tradescantia zebrina* (Purple Wandering Jew)

Succulents

Note: Many of these plants are also tropical or subtropical.

- *Aloe vera*
- Cactaceae (Cacti)
    - *Epiphyllum* (Orchid Cacti)
    - *Mammillaria*
    - *Opuntia* (Large genus that includes the Prickly Pear)
    - *Zygocactus* (Christmas Cactus)
- *Crassula ovata* (Jade Plant)

Forced Bulbs

Note: Many forced bulbs are also temperate.

- *Crocus*
- *Hyacinthus* (Hyacinth)
- *Narcissus (genus)* (Narcissus or Daffodil)

Temperates

- *Hedera helix* (English Ivy)
- *Saxifraga stolonifera* (Strawberry Begonia)

Urban Horticulture

Horticulture is the science and art of growing fruits and vegetables and also flowers or ornamental plants.

Urban horticulture specifically is the study of the relationship between plants and the urban environment. It focuses on the functional use of horticulture so as to maintain and improve the surrounding urban area. With the expansion of cities and rapid urbanization, urban horticulture is a new term. As a result of the urban context, this field of study is large and complex and its study has only recently gained momentum. It has an undeniable relationship to production horticulture in that fruits, vegetables and other plants are grown for harvest, aesthetic, architectural, recreational and psychological purposes, but it extends far beyond these benefits. The value of plants in the urban environment has yet to be thoroughly researched or quantified.

Salad lettuce cultivation at the Growing Communities' urban plot, in Springfield Park, Clapton, North London.

A variety of flowers and vegetables grown under metal halide lamps

History

Horticulture and the integration of nature into our civilization has been a major part in the establishment of our cities. When nomadic civilizations began settling down, their major trading centers were the market gardens and farms. Urban horticulture rapidly progressed with the birth of cities and the increase in experimentation and exchange of ideas. These insights led to the field being dispersed to farmers in the hinterlands. For centuries, the built environment such as homes, public buildings, etc. were integrated with cultivation in the form of gardens, farms, and grazing lands, Kitchen gardens, farms, common grazing land, etc. Therefore, horticulture was a regular part of everyday life in the city. With the Industrial Revolution and the related increasing populations rapidly changed the landscape and replaced green spaces with brick and asphalt. After the nineteenth century, Horticulture was then selectively restored in some urban spaces as a response to the unhealthy conditions of factory neighborhoods and cities began seeing the development of parks.

Post World War II Trends

Early urban horticulture movements majorly served the purposes of short term welfare during recession periods, philanthropic charity to uplift "the masses" or patriotic relief. The tradition of urban horticulture mostly declined after World War II as suburbs became the focus of residential and commercial growth. Most of the economically stable population moved out of the cities into the suburbs, leaving only slums and ghettos at the city centers. However, there were a few exceptions of garden projects initiated by public housing authorities in the 1950s and 60s for the purpose of beautification and tenant pride. But for the most part as businesses also left the metropolitan areas, it generated wastelands and areas of segregated poverty.

Inevitably the disinvestment of major city centers, specifically in America resulted in the drastic increase of vacant lots. Existing buildings became uninhabitable, houses were abandoned and even productive industrial land became vacant. Modern community gardening, urban agriculture, and food security movements were a form of response to battle the above problems at a local level. In fact other movements at that time such as the peace, environmental, women's, civil rights, and "back-to-the-city" movements of the 1960s and 70's and the environmental justice movement of the 1980s and 90's saw opportunity in these vacant lands as a way of reviving communities through school and community gardens, farmers' markets, and urban agriculture.

Modern Community Garden Movement

Things have taken a turn in the twenty-first century as people are recognizing the need for local community gardens and green spaces. It is not the concept but the purposes that are new. The main goals of this movement include cleaning up neighborhoods, pushing out drug dealing that occurs at empty lots, growing and preserving food for consumption, restoring nature to industrial areas, and bringing the farming traditions to urban cities. Essentially community gardening is seen as way of creating a relationship between people and a place through social and physical engagement. Most urban gardens are created on vacant land that vary in size and are generally gardened as individual plots by community members. Such areas can support social, cultural, and artistic events and contribute to the rebuilding of local community spirit. The modern community garden movement is initiated by neighborhoods along with the support of the governments and non-profit organizations. Some gardens are linked to public housing projects, schools through garden-based learning programs, churches and social agencies and some even employ those who are incarcerated. Community gardens which are now a large part of the urban horticulture movement are different from the earlier periods of grand park development in that the latter only served to free the people from the industrialism. In addition a community garden is more beneficial and engaging than a mere lawn or park and serves as a valuable access to nature where wilderness is unavailable. This movement helped create and sustain relationships between city dwellers and the soil and contributed to a different kind of urban environmentalism that did not have any characteristics of reform charity.

Despite that it has been 30 years since the first community gardens in the States, there is no concrete analysis of current urban gardens and their organizations. The American Community Gardening Association (ACGA) has estimations that show that municipal governments and non-profit organizations operate gardening programs in about 250 cities and towns, although the staff of this organization admits that this number could in reality be twice as large. In 1994 survey, the National Gardening Association found that 6.7 million households, that weren't involved in gardening would be interested in doing so if there was a plot nearby. A more recent survey showed that more gardens are being created in cities as opposed to being lost to economic development.

Today urban horticulture has several components that include more than just community gardens, such as market gardens, small farms and farmers' markets and is an important aspect of community development. Another result of urban horticulture is the food security movement where locally grown food is given precedence through several projects and programs, thus providing low-cost and nutritious food. Urban community gardens and the food security movement was a response to the problems of industrial agriculture and to solve its related problems of price inflation, lack of supermarkets, food scarcity, etc.

Benefits

Horticulture by itself is a practical and applied science, which means it can have a significance in our everyday lives. As community gardens cannot actually compete with market-based land uses, it is essential to find other ways to understand their various benefits such as their contribution to social, human, and financial well-being. Frederick Law Olmsted, the designer of New York City's Central Park observed that the trees, meadows, ponds and wildlife tranquilize the stresses of city life. According to various studies over the years, nature has a very positive impact over human health and even more so in an emotional and psychological sense. Trees, grass, and flower gardens, due to their presence as well as visibility, increase people's life satisfaction by reducing fatigue and irritation and restoring a sense of calm. In fact Honeyman tested the restorative value of nature scenes in urban settings and discovered that vegetation in an urban setting produced more mental restoration as opposed to areas without vegetation. In addition, areas with only nature did not have as much of a positive psychological impact as did the combination of urban areas and nature.

One of the obvious health benefits of gardening is the increased intake of fruits and vegetables. But the act of gardening itself, is also a major health benefit. Gardening is a low-impact exercise, which when added into daily activities, can help reduce weight, lower stress, and improve overall health. A recent study showed a reduced body mass index and lower weight in community gardeners compared with their non-gardening counterparts The study showed men who gardened had a body mass index -2.36 lower and were 62% less likely to be overweight than their neighbors, while women were 46% less likely to be overweight with a body mass index -1.88 lower than their neighbors. Access to urban gardens can improve health through nutritious, edible plantings, as well by getting people outside and promoting more activity in their environments.

Gardening programs in inner-city schools have become increasingly popular as a way to teach children not only about healthy eating habits, but also to encourage students to become active learners. Besides getting students outside and moving, and encouraging an active lifestyle, children also learn leadership, teamwork, communication and collaboration skills, in addition to critical and creative thinking skills. Gardening in schools will enable children to share with their families the health and nutrition benefits of eating fresh fruits and vegetables. Because weather and soil conditions are in a state of constant change, students learn to adapt their thinking and creatively problem solve, depending on the situations that arise. Students also learn to interact and communicate with a diverse population of people, from other students to adult volunteers. These programs benefit students' health and enable them to be active contributors in the world around them.

Gardens and other green spaces also increase social activity and help in creating a sense of place, apart from their various other purposes such as enhancing the community by mediating environmental factors. There is also a huge disparity in the availability of sources that provide nutritious and affordable foods especially around urban centers which have problems of poverty, lack of public transport and abandonment by supermarkets. Therefore, inner city community gardens can be a valuable sources of nutrition at an affordable cost in the most easily accessible way.

In order to understand and thereby maximize the benefits of urban horticulture, it is essential to document the effects of horticulture activities and quantify the benefits so that governments and private industries can make the appropriate changes. Horticulturists have always been involved in the botanical and physical aspects of horticulture but an involvement in its social and emotional factors would be highly beneficial to communities, cities and to the field of horticulture and its profession. Based on this, in the 1970s, the International Society for Horticultural Science recognized this need for research on the functional use of plants in an urban setting along with the need of improved communication between scientists in this field of research and people who utilize plants. The Commission for Urban Horticulture was established in 1982 which deals with plants grown in urban areas, management techniques, the functional use of these plants as well the shortcomings of the current lack of knowledge regarding this field. The establishment of such a commission is an important indicator that this topic has reached a level of international recognition.

Economic Benefits

There are many different economic benefits from gardening from saving money purchasing food and even on the utility bills. Developing countries can spend up to 60-80 percent of income on buying food alone. In Barbara Lake, Milfront Taciano and Gavin Michaels Journal of Psychology title "The Relative Influence of Psycho-Social Factors on Urban Gardening," they say that while people are saving money on buying food, having roof top gardens are also becoming popular. Having green roofs can reduce the cost of heating in the winter and help stay cool in the summer. Green roofs also can lower the cost of roof replacement. While green roofs are an addition to urban horticulture people are eating healthy while also improving the value of their property. Other benefits include increased employment from non-commercial jobs where producers include reductions on the cost of food, (Lake, Taciano, and Michael).

Production Practices

Crops are grown in flowerpots, growbags, small gardens or larger fields, using traditional or high-tech and innovative practices. Some new techniques that have been adapted to the urban situation and tackle the main city restrictions are also documented. These include horticultural production on built-up land using various types of substrates (e.g. roof top, organic production and hydroponic/aeroponic production). Because of this, it is also known as roof-top vegetable gardennig/ horticulture and container vegetable gardening/horticulture.

Urban and Peri-urban Horticulture in Africa

A report of the United Nations Food and Agriculture Organization, *Growing greener cities in Africa*, states that market gardening – i.e. irrigated, commercial production of fruit and vegetables in areas designated for the purpose, or in other urban open spaces – is the single most important source of locally grown, fresh produce in 10 out of 27 African countries for which data are available. Market gardening produces most of all the leafy vegetables consumed in Accra, Dakar, Bangui, Brazzaville, Ibadan, Kinshasa and Yaoundé, cities that, between them, have a total population of 22.5 million. Market gardens provide around half of the leafy vegetable supply in Addis Ababa, Bissau and Libreville. The report says that in most of urban Africa, market gardening is an informal and often illegal activity, which has grown with little official recognition, regulation or support. Most gardeners have no formal title to their land, and many lose it overnight. Land suitable for horticulture is being taken for housing, industry and infrastructure. To maximize earnings from insecure livelihoods, many gardeners are overusing pesticide and urban waste water.

Plug (Horticulture)

Plugs in horticulture are small-sized seedlings grown in trays from expanded polystyrene or polythene filled usually with a peat or compost substrate. This type of plug is used for commercially raising vegetables and bedding plants. Similarly plugs may also refer to small sections of lawn grass sod. After being planted, lawn grass may somewhat spread over an adjacent area.

Plugs of yellow pear tomatoes

Plug plants are young plants raised in small, individual cells, ready to be transplanted into containers or a garden. Professionally raised vegetable/flowering plants in controlled conditions during their important formative period (the first 4–6 weeks) can help to ensure plant health and for plants to reach their maximum potential during the harvest/blooming period. Establishing a garden using plug plants is often easier than doing so starting from seed.

Overview

Plugs growing in a greenhouse at New Place Nurseries in Pulborough, West Sussex, England

Planting from plugs reduces the time a crop resides in the ground, and is functional for those with limited space. Plugs can improve yields: a healthy, stocky plant will grow rapidly and symmetrically when planted out, with a potentially greater capacity to withstand pests, disease and drought. Raising some types of seedlings successfully can be difficult, so plug plants can be beneficial for less experienced gardeners. Plug plants are beneficial for gardeners who want to try a new variety or a range of varieties without purchasing numerous packets of seeds and starting the plants from seed. Plug plants are very useful if the sowing window is missed, and plugs can be purchased quickly to replace a crop which has failed.

As a garden develops, interplanting (intercropping) existing crops with plugs plants, ideally companion plants, can improve the productivity of the space and so maximise harvests – a sown crop may not be able to compete with established plants. Plug plants are much easier to weed than sown seedlings, and weeding will need to be done less frequently.

Having semi-grown plants simplifies designing a vegetable plot or container. As plants that have already started growth, the time to attain plant growth is lessened. Within days of planting signs of growth are typically visible: leaves will perk up and roots anchor into the soil. Air pruned plugs are grown in a manner to promote very rapid growth almost immediately after being transplanted to new soil.

Plugs are sometimes used in hillside plasticulture applications, due to the ease in which they are transplanted.

Plant Cultivation and Growth

A plug of St. Augustine grass ready for sprigging

Plug plants grow more consistently, as has been noted by the commercial scale vegetable growing industry, and more rapidly; large-scale brassica field crops are planted almost exclusively from soil block plugs in some parts of Europe, a trend which is growing in the UK. This success at the commercial scale is testament to the success of plugs in the ground.

It is of note that many varieties actively benefit from being transplanted as severing the taproot encourages bushier root growth. Traditionally nearly all heading brassica are sown in a separate seed bed, thinned, and the best ones planted in a prepared bed after about 6–8 weeks. Many pests want to eat baby brassica; this in combination with its long growing season makes planting brassica from plugs a much easier option.

Root Vegetables

Root vegetables are typically, but not always, sown from seed, rather than transplanted from plugs, where they are to mature and then be thinned. The thinning action is highly beneficial in itself as it provides soil aeration at depth without disturbing adjacent roots systems. The initial concentration of seedlings also dilutes damage from pests and provided some food for the gardener or the compost in the form of thinnings. Beetroot, carrots and the root brassica family- swede, turnip- will simply not reach their full potential with any check to early root growth. In addition, these seeds are typically inexpensive, and the seedlings are delicate; hence there is little value to the gardener in buying or growing them as plugs.

References

- Spencer, Roger, ed. (1995–2000), Horticultural Flora of South-eastern Australia: The Identification of Garden & Cultivated Plants, Sydney, Australia: University of New South Wales Press, ISBN 0 86840 167 6 (set).
- Staples, George W; Herbst, Derral R (2005), A Tropical Garden Flora: Plants cultivated in the Hawaiian Islands and other tropical places, Honolulu: Bishop Museum Press/Mutual Pub Co, ISBN 1581780397.
- Rehder, Alfred (1987), Manual of Cultivated Trees and Shrubs Hardy in North America (New edition of Second Revised ed.), Portland, Oregon: Timber Press, ISBN 0-931146-00-3.
- Robinson, Muriel. "Urban Horticulture- Purpose and Prospects". The Robinson Garden at Earlscliffe, Baily, Co. Dublin, Ireland. Retrieved 6 November 2014.
- Lake, Barbara, Taciano L. Milfront C. Gavin. "The Relative Influence of Psycho-Social Factors on Urban Gardening." Academic Search Premier. Web. 10 June 2013.
- Dunnett, Nigel; et al. (2011). Small Green Roofs: Low-Tech Options for Greener Living. Timber Press. pp. 67– 68. Retrieved March 30, 2012. ISBN 9781604690590.
- Greenwood, Pippa; Beazley, Mitchell (2008). 1001 Ideas for Better Gardening. Creative Homeowner Press. p. 106. Retrieved March 30, 2012. ISBN 9781580111805.

6

Study of Plant Pathology

The study of diseases caused by pathogens and harmful ecological conditions is known as plant pathology. This chapter introduces the reader to plant pathology and lists the various pathogens, pests and conditions that lead to infection in plants. The text lists several well-known plant diseases, their symptoms and epidemiology. The topic of disease resistance and the various mechanisms of resistance are tackled in this chapter. Horticulture is best understood in confluence with the major topics listed in the following chapter.

Plant Pathology

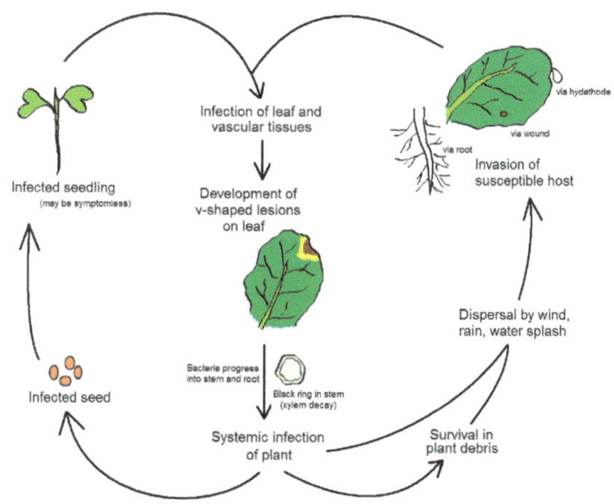

Life cycle of the black rot pathogen, *Xanthomonas campestris* pathovar *campes*

Plant pathology (also phytopathology) is the scientific study of diseases in plants caused by pathogens (infectious organisms) and environmental conditions (physiological factors). Organisms that cause infectious disease include fungi, oomycetes, bacteria, viruses, viroids, virus-like organisms, phytoplasmas, protozoa, nematodes and parasitic plants. Not included are ectoparasites like insects, mites, vertebrate, or other pests that affect plant health by consumption of plant tissues. Plant pathology also involves the study of pathogen identification, disease etiology, disease cycles, economic impact, plant disease epidemiology, plant disease resistance, how plant diseases affect humans and animals, pathosystem genetics, and management of plant diseases.

Overview

Control of plant diseases is crucial to the reliable production of food, and it provides significant reductions in agricultural use of land, water, fuel and other inputs. Plants in both natural and cultivated populations carry inherent disease resistance, but there are numerous examples of devastating plant disease impacts, as well as recurrent severe plant diseases. However, disease control is reasonably successful for most crops. Disease control is achieved by use of plants that have been bred for good resistance to many diseases, and by plant cultivation approaches such as crop rotation, use of pathogen-free seed, appropriate planting date and plant density, control of field moisture, and pesticide use. Across large regions and many crop species, it is estimated that diseases typically reduce plant yields by 10% every year in more developed settings, but yield loss to diseases often exceeds 20% in less developed settings. Continuing advances in the science of plant pathology are needed to improve disease control, and to keep up with changes in disease pressure caused by the ongoing evolution and movement of plant pathogens and by changes in agricultural practices. Plant diseases cause major economic losses for farmers worldwide. The Food and Agriculture Organization estimates indeed that pests and diseases are responsible for about 25% of crop loss. To solve this issue, new methods are needed to detect diseases and pests early, such as novel sensors that detect plant odours and spectroscopy and biophotonics that are able to diagnostic plant health and metabolism.

Plant Pathogens

Fungi

Most phytopathogenic fungi belong to the Ascomycetes and the Basidiomycetes.

The fungi reproduce both sexually and asexually via the production of spores and other structures. Spores may be spread long distances by air or water, or they may be soilborne. Many soil inhabiting fungi are capable of living saprotrophically, carrying out the part of their life cycle in the soil. These are known as facultative saprotrophs.

Powdery mildew, a biotrophic fungus

Fungal diseases may be controlled through the use of fungicides and other agriculture practices. However, new races of fungi often evolve that are resistant to various fungicides.

Biotrophic fungal pathogens colonize living plant tissue and obtain nutrients from living host cells. Necrotrophic fungal pathogens infect and kill host tissue and extract nutrients from the dead host cells.

Rice blast, caused by a necrotrophic fungus

Significant fungal plant pathogens include:

Ascomycetes

- *Fusarium* spp. (causal agents of Fusarium wilt disease)
- *Thielaviopsis* spp. (causal agents of: canker rot, black root rot, *Thielaviopsis* root rot)
- *Verticillium* spp.
- *Magnaporthe grisea* (causal agent of rice blast)
- *Sclerotinia sclerotiorum* (causal agent of cottony rot)

Basidiomycetes

- *Ustilago* spp. (causal agents of smut)
- *Rhizoctonia* spp.
- *Phakospora pachyrhizi* (causal agent of soybean rust)
- *Puccinia* spp. (causal agents of severe rusts of virtually all cereal grains and cultivated grasses)
- *Armillaria* spp. (the so-called honey fungus species, which are virulent pathogens of trees and produce edible mushrooms)

Fungus-like Organisms

Oomycetes

The oomycetes are not true fungi but are fungus-like organisms. They include some of the most destructive plant pathogens including the genus *Phytophthora*, which includes the causal agents of potato late blight and sudden oak death. Particular species of oomycetes are responsible for root rot.

Despite not being closely related to the fungi, the oomycetes have developed very similar infection strategies. Oomycetes are capable of using effector proteins to turn off a plant's defenses in its infection process. Plant pathologists commonly group them with fungal pathogens.

Significant oomycete plant pathogens

- *Pythium* spp.
- *Phytophthora* spp., including the causal agent of the Great Irish Famine (1845–1849)

Phytomyxea

Some slime molds in Phytomyxea cause important diseases, including club root in cabbage and its relatives and powdery scab in potatoes. These are caused by species of *Plasmodiophora* and *Spongospora*, respectively.

Bacteria

Most bacteria that are associated with plants are actually saprotrophic and do no harm to the plant itself. However, a small number, around 100 known species, are able to cause disease. Bacterial diseases are much more prevalent in subtropical and tropical regions of the world.

Most plant pathogenic bacteria are rod-shaped (bacilli). In order to be able to colonize the plant they have specific pathogenicity factors. Five main types of bacterial pathogenicity factors are known: uses of cell wall–degrading enzymes, toxins, effector proteins, phytohormones and exopolysaccharides.

Pathogens such as *Erwinia* species use cell wall–degrading enzymes to cause soft rot. *Agrobacterium* species change the level of auxins to cause tumours with phytohormones. Exopolysaccharides are produced by bacteria and block xylem vessels, often leading to the death of the plant.

Bacteria control the production of pathogenicity factors via quorum sensing.

Vitis vinifera with "Ca. Phytoplasma vitis" infection

Significant bacterial plant pathogens:

- Burkholderia
- Proteobacteria
    - *Xanthomonas* spp.
    - *Pseudomonas* spp.

- Pseudomonas syringae pv. tomato causes tomato plants to produce less fruit, and it "continues to adapt to the tomato by minimizing its recognition by the tomato immune system."

Phytoplasmas ('Mycoplasma-like Organisms') and Spiroplasmas

Phytoplasma and *Spiroplasma* are a genre of bacteria that lack cell walls and are related to the mycoplasmas, which are human pathogens. Together they are referred to as the mollicutes. They also tend to have smaller genomes than most other bacteria. They are normally transmitted by sap-sucking insects, being transferred into the plants phloem where it reproduces.

Tobacco mosaic virus

Viruses, Viroids and Virus-like Organisms

There are many types of plant virus, and some are even asymptomatic. Under normal circumstances, plant viruses cause only a loss of crop yield. Therefore, it is not economically viable to try to control them, the exception being when they infect perennial species, such as fruit trees.

Most plant viruses have small, single-stranded RNA genomes. However some plant viruses also have double stranded RNA or single or double stranded DNA genomes. These genomes may encode only three or four proteins: a replicase, a coat protein, a movement protein, in order to allow cell to cell movement through plasmodesmata, and sometimes a protein that allows transmission by a vector. Plant viruses can have several more proteins and employ many different molecular translation methods.

Plant viruses are generally transmitted from plant to plant by a vector, but mechanical and seed transmission also occur. Vector transmission is often by an insect (for example, aphids), but some fungi, nematodes, and protozoa have been shown to be viral vectors. In many cases, the insect and virus are specific for virus transmission such as the beet leafhopper that transmits the curly top virus causing disease in several crop plants.

Nematodes

Nematodes are small, multicellular wormlike animals. Many live freely in the soil, but there are some species that parasitize plant roots. They are a problem in tropical and subtropical regions of the world, where they may infect crops. Potato cyst nematodes (*Globodera pallida* and *G. rosto-*

chiensis) are widely distributed in Europe and North and South America and cause $300 million worth of damage in Europe every year. Root knot nematodes have quite a large host range, whereas cyst nematodes tend to be able to infect only a few species. Nematodes are able to cause radical changes in root cells in order to facilitate their lifestyle.

Root-knot nematode galls

Protozoa and Algae

There are a few examples of plant diseases caused by protozoa (e.g., *Phytomonas*, a kinetoplastid). They are transmitted as zoospores that are very durable, and may be able to survive in a resting state in the soil for many years. They have also been shown to transmit plant viruses.

When the motile zoospores come into contact with a root hair they produce a plasmodium and invade the roots.

Some colourless parasitic algae (e.g., *Cephaleuros*) also cause plant diseases.

Parasitic Plants

Parasitic plants such as mistletoe and dodder are included in the study of phytopathology. Dodder, for example, is used as a conduit either for the transmission of viruses or virus-like agents from a host plant to a plant that is not typically a host or for an agent that is not graft-transmissible.

Common Pathogenic Infection Methods

- Cell wall-degrading enzymes: These are used to break down the plant cell wall in order to release the nutrients inside.
- Toxins: These can be non-host-specific, which damage all plants, or host-specific, which cause damage only on a host plant.
- Effector proteins: These can be secreted into the extracellular environment or directly into the host cell, often via the Type three secretion system. Some effectors are known to suppress host defense processes. This can include: reducing the plants internal signaling mechanisms or reduction of phytochemicals production. Bacteria, fungus and oomycetes are known for this function.

Physiological Plant Disorders

Significant abiotic disorders can be caused by:

Natural

Drought

Frost damage and breakage by snow and hail

Flooding and poor drainage

Nutrient deficiency

Salt deposition and other soluble mineral excesses (e.g., gypsum)

Wind (windburn and breakage by hurricanes and tornadoes)

Lightning and wildfire (also often man-made)

Man-made (arguably not abiotic, but usually regarded as such)

Soil compaction

Pollution of air, soil, or both

Salt from winter road salt application or irrigation

Herbicide over-application

Poor education and training of people working with plants (e.g. lawnmower damage to trees)

Vandalism

Orchid leaves with viral infections

Epidemiology

The study of phytopathology epidemics

Disease Resistance

Management

Quarantine

A diseased patch of vegetation or individual plants can be isolated from other, healthy growth. Specimens may be destroyed or relocated into a greenhouse for treatment or study. Another option is to avoid the introduction of harmful nonnative organisms by controlling all human traffic and activity (e.g., AQIS), although legislation and enforcement are crucial in order to ensure lasting effectiveness.

Cultural

Farming in some societies is kept on a small scale, tended by peoples whose culture includes farming traditions going back to ancient times. (An example of such traditions would be lifelong training in techniques of plot terracing, weather anticipation and response, fertilization, grafting, seed care, and dedicated gardening.) Plants that are intently monitored often benefit from not only active external protection but also a greater overall vigor. While primitive in the sense of being the most labor-intensive solution by far, where practical or necessary it is more than adequate.

Plant resistance

Sophisticated agricultural developments now allow growers to choose from among systematically cross-bred species to ensure the greatest hardiness in their crops, as suited for a particular region's pathological profile. Breeding practices have been perfected over centuries, but with the advent of genetic manipulation even finer control of a crop's immunity traits is possible. The engineering of food plants may be less rewarding, however, as higher output is frequently offset by popular suspicion and negative opinion about this "tampering" with nature.

Chemical

Many natural and synthetic compounds can be employed to combat the above threats. This method works by directly eliminating disease-causing organisms or curbing their spread; however, it has been shown to have too broad an effect, typically, to be good for the local ecosystem. From an economic standpoint, all but the sim-plest natural additives may disqualify a product from "organic" status, potentially reducing the value of the yield.

Biological

Crop rotation may be an effective means to prevent a parasitic population from becoming well-established, as an organism affecting leaves would be starved when the leafy crop is replaced by a tuberous type, etc. Other means to undermine parasites without attacking them directly may exist.

Integrated

The use of two or more of these methods in combination offers a higher chance of effectiveness.

Timeline of Plant Pathology

300–286 BC Theophrastus, father of botany, wrote and studied diseases of trees, cereals and legumes

1665 Robert Hooke illustrates a plant-pathogenic fungal disease, rose rust

1675 Anton van Leeuwenhouek invents the compound microscope, in 1683 describes bacteria seen with the microscope

1729 Pier Antonio Micheli, father of mycology, observes spores for the first time, conducts germination experiments

1755 Tillet reports on treatment of seeds

1802 Lime sulfur first used to control plant disease

1845–1849 Potato late blight epidemic in Ireland

1853 Heinrich Anton de Bary father of modern mycology, establishes that fungi are the cause, not the result, of plant diseases, publishes "Untersuchungen uber die Brandpilze"

1858 Julius Kühn publishes "Die Krankheiten der Kultergewachse"

1865 M. Planchon discovers a new species of *Phylloxera*, which was named *Phylloxera vastatrix.*

1868–1882 Coffee rust epidemic in Sri Lanka

1875 Mikhail Woronin identified the cause of clubroot as a "plasmodiophorous organism" and gave it the name *Plasmodiophora brassicae*

1876 *Fusarium oxysporum* f.sp. *cubense*, responsible for Panama disease, discovered in bananas in Australia

1878–1885 Downy mildew of grape epidemic in France

1879 Robert Koch establishes germ theory: diseases are caused by microorganisms

1882 *Lehrbuch der Baumkrankheiten* (*Textbook of Diseases of Trees*), by Robert Hartig, is published in Berlin, the first textbook of forest pathology.

1885 Bordeaux mixture introduced by Pierre-Marie-Alexis Millardet to control downy mildew on grape

1885 Experimental proof that bacteria can cause plant diseases: "Erwinia amylovora" and fire blight of apple

1886–1898 Recognition of plant viral diseases: Tobacco mosaic virus

1889 Introduction of hot water treatment of seed for disease control by Jensen

1902 First chair of plant pathology established, in Copenhagen

1904 Mendelian inheritance of cereal rust resistance demonstrated

1907 First academic department of plant pathology established, at Cornell University

1908 American Phytopathological Society founded

1910 Panama disease reaches Western Hemisphere

1911 Scientific journal *Phytopathology* founded

1925 Panama disease reaches every banana-growing country in the Western Hemisphere

1951 European and Mediterranean Plant Protection Organization (EPPO) founded

1967 Recognition of plant pathogenic mycoplasma-like organisms

1971 T. O. Diener discovers viroids, organisms smaller than viruses

The historical landmarks in plant pathology are taken from unless otherwise noted.

Plant Virus

Plant viruses are viruses that affect plants. Like all other viruses, plant viruses are obligate intracellular parasites that do not have the molecular machinery to replicate without a host. Plant viruses are pathogenic to higher plants. While this article does not intend to list all plant viruses, it discusses some important viruses as well as their uses in plant molecular biology.

Pepper mild mottle virus

Leaf curl virus

Overview

Although plant viruses are not nearly as well understood as the animal counterparts, one plant virus has become iconic. The first virus to be discovered was *Tobacco mosaic virus* (TMV). This and other viruses cause an estimated US$60 billion loss in crop yields worldwide each year. Plant viruses are grouped into 73 genera and 49 families. However, these figures relate only to cultivated plants that represent only a tiny fraction of the total number of plant species. Viruses in wild plants have been poorly studied, but those studies that exist almost overwhelming show that such interactions between wild plants and their viruses do not appear to cause disease in the host plants.

To transmit from one plant to another and from one plant cell to another, plant viruses must use strategies that are usually different from animal viruses. Plants do not move, and so plant-to-plant transmission usually involves vectors (such as insects). Plant cells are surrounded by solid cell walls, therefore transport through plasmodesmata is the preferred path for virions to move between plant cells. Plants probably have specialized mechanisms for transporting mRNAs through plasmodesmata, and these mechanisms are thought to be used by RNA viruses to spread from one cell to another.

Plant defenses against viral infection include, among other measures, the use of siRNA in response to dsRNA. Most plant viruses encode a protein to suppress this response. Plants also reduce transport through plasmodesmata in response to injury.

History

The discovery of plant viruses causing disease is often accredited to A. Mayer (1886) working in the Netherlands demonstrated that the sap of mosaic obtained from tobacco leaves developed mosaic symptom when injected in healthy plants. However the infection of the sap was destroyed when it was boiled. He thought that the causal agent was the bacteria. However, after larger inoculation with a large number of bacteria, he failed to develop a mosaic symptom.

In 1898, Martinus Beijerinck, who was a Professor of Microbiology at the Technical University the Netherlands, put forth his concepts that viruses were small and determined that the "mosaic disease" remained infectious when passed through a Chamberland filter-candle. This was in contrast to bacteria microorganisms, which were retained by the filter. Beijerinck referred to the infectious filtrate as a "contagium vivum fluidum", thus the coinage of the modern term "virus".

After the initial discovery of the 'viral concept' there was need to classify any other known viral diseases based on the mode of transmission even though microscopic observation proved fruitless. In 1939 Holmes published a classification list of 129 plant viruses. This was expanded and in 1999 there were 977 officially recognized, and some provisional, plant virus species.

The purification (crystallization) of TMV was first performed by Wendell Stanley, who published his findings in 1935, although he did not determine that the RNA was the infectious material. However, he received the Nobel Prize in Chemistry in 1946. In the 1950s a discovery by two labs simultaneously proved that the purified RNA of the TMV was infectious which reinforced the argument. The RNA carries genetic information to code for the production of new infectious particles.

More recently virus research has been focused on understanding the genetics and molecular biology of plant virus genomes, with a particular interest in determining how the virus can replicate, move and infect plants. Understanding the virus genetics and protein functions has been used to explore the potential for commercial use by biotechnology companies. In particular, viral-derived sequences have been used to provide an understanding of novel forms of resistance. The recent boom in technology allowing humans to manipulate plant viruses may provide new strategies for production of value-added proteins in plants.

Structure

Viruses are extremely small and can only be observed under an electron microscope. The structure of a virus is given by its coat of proteins, which surround the viral genome. Assembly of viral particles takes place spontaneously.

Over 50% of known plant viruses are rod-shaped (flexuous or rigid). The length of the particle is normally dependent on the genome but it is usually between 300–500 nm with a diameter of 15–20 nm. Protein subunits can be placed around the circumference of a circle to form a disc. In the presence of the viral genome, the discs are stacked, then a tube is created with room for the nucleic acid genome in the middle.

The second most common structure amongst plant viruses are isometric particles. They are 25–50 nm in diameter. In cases when there is only a single coat protein, the basic structure consists of 60 T subunits, where T is an integer. Some viruses may have 2 coat proteins that associate to form an icosahedral shaped particle.

There are three genera of *Geminiviridae* that possess geminate particles which are like two isometric particles stuck together.

A very small number of plant viruses have, in addition to their coat proteins, a lipid envelope. This is derived from the plant cell membrane as the virus particle buds off from the cell.

Transmission of Plant Viruses

Through Sap

Viruses can be spread by direct transfer of sap by contact of a wounded plant with a healthy one. Such contact may occur during agricultural practices, as by damage caused by tools or hands, or naturally, as by an animal feeding on the plant. Generally TMV, potato viruses and cucumber mosaic viruses are transmitted via sap.

Insects

Plant viruses need to be transmitted by a vector, most often insects such as leafhoppers. One class of viruses, the Rhabdoviridae, has been proposed to actually be insect viruses that have evolved to replicate in plants. The chosen insect vector of a plant virus will often be the determining factor in that virus's host range: it can only infect plants that the insect vector feeds upon. This was shown in part when the old world white fly made it to the United States, where it transferred many plant viruses into new hosts. Depending on the way they are transmitted, plant viruses are classified as non-persistent, semi-persistent and persistent. In non-persistent transmission, viruses become attached to the distal tip of the stylet of the insect and on the next plant it feeds on, it inoculates it with the virus. Semi-persistent viral transmission involves the virus entering the foregut of the insect. Those viruses that manage to pass through the gut into the haemolymph and then to the salivary glands are known as persistent. There are two sub-classes of persistent viruses: propagative and circulative. Propagative viruses are able to replicate in both the plant and the insect (and may have originally been insect viruses), whereas circulative can not. Circulative viruses are protected inside aphids by the chaperone protein symbionin, produced by bacterial symbionts. Many plant viruses encode within their genome polypeptides with domains essential for transmission by insects. In non-persistent and semi-persistent viruses, these domains are in the coat protein and another protein known as the helper component. A bridging hypothesis has been proposed to explain how these proteins aid in insect-mediated viral transmission. The helper component will bind to the specific domain of the coat protein, and then the insect mouthparts — creating a bridge. In persistent propagative viruses, such as tomato spotted wilt virus (TSWV), there is often a lipid coat surrounding the proteins that is not seen in other classes of plant viruses. In the case of TSWV, 2 viral proteins are expressed in this lipid envelope. It has been proposed that the viruses bind via these proteins and are then taken into the insect cell by receptor-mediated endocytosis.

Nematodes

Soil-borne nematodes also have been shown to transmit viruses. They acquire and transmit them by feeding on infected roots. Viruses can be transmitted both non-persistently and persistently, but there is no evidence of viruses being able to replicate in nematodes. The virions attach to the stylet (feeding organ) or to the gut when they feed on an infected plant and can then detach during later feeding to infect other plants. Examples of viruses that can be transmitted by nematodes include tobacco ringspot virus and tobacco rattle virus.

Plasmodiophorids

A number of virus genera are transmitted, both persistently and non-persistently, by soil borne zoosporic protozoa. These protozoa are not phytopathogenic themselves, but parasitic. Transmission of the virus takes place when they become associated with the plant roots. Examples include *Polymyxa graminis*, which has been shown to transmit plant viral diseases in cereal crops and *Polymyxa betae* which transmits Beet necrotic yellow vein virus. Plasmodiophorids also create wounds in the plant's root through which other viruses can enter.

Seed and Pollen Borne Viruses

Plant virus transmission from generation to generation occurs in about 20% of plant viruses. When viruses are transmitted by seeds, the seed is infected in the generative cells and the virus is maintained in the germ cells and sometimes, but less often, in the seed coat. When the growth and development of plants is delayed because of situations like unfavourable weather, there is an increase in the amount of virus infections in seeds. There does not seem to be a correlation between the location of the seed on the plant and its chances of being infected. Little is known about the mechanisms involved in the transmission of plant viruses via seeds, although it is known that it is environmentally influenced and that seed transmission occurs because of a direct invasion of the embryo via the ovule or by an indirect route with an attack on the embryo mediated by infected gametes. These processes can occur concurrently or separately depending on the host plant. It is unknown how the virus is able to directly invade and cross the embryo and boundary between the parental and progeny generations in the ovule. Many plants species can be infected through seeds including but not limited to the families Leguminosae, Solanaceae, Compositae, Rosaceae, Cucurbitaceae, Gramineae. Bean common mosaic virus is transmitted through seeds.

Direct Plant-to-human Transmission

Researchers from the University of the Mediterranean in Marseille, France have found tenuous evidence that suggest a virus common to peppers, the Pepper Mild Mottle Virus (PMMoV) may have moved on to infect humans. This is a very rare and highly unlikely event as, to enter a cell and replicate, a virus must "bind to a receptor on its surface, and a plant virus would be highly unlikely to recognize a receptor on a human cell. One possibility is that the virus does not infect human cells directly. Instead, the naked viral RNA may alter the function of the cells through a mechanism similar to RNA interference, in which the presence of certain RNA sequences can turn genes on and off," according to Virologist Robert Garry from the Tulane University in New Orleans, Louisiana.

Translation of Plant Viral Proteins

75% of plant viruses have genomes that consist of single stranded RNA (ssRNA). 65% of plant viruses have +ssRNA, meaning that they are in the same sense orientation as messenger RNA but 10% have -ssRNA, meaning they must be converted to +ssRNA before they can be translated. 5% are double stranded RNA and so can be immediately translated as +ssRNA viruses. 3% require a reverse transcriptase enzyme to convert between RNA and DNA. 17% of plant viruses are ssDNA and very few are dsDNA, in contrast a quarter of animal viruses are dsDNA and three quarters of bacteriophage are dsDNA. Viruses use the plant ribosomes to produce the 4-10 proteins encoded by their genome. However, since many of the proteins are encoded on a single strand (that is, they are polycistronic) this will mean that the ribosome will either only produce one protein, as it will terminate translation at the first stop codon, or that a polyprotein will be produced. Plant viruses have had to evolve special techniques to allow the production of viral proteins by plant cells.

5' Cap

For translation to occur, eukaryotic mRNAs require a 5' Cap structure. This means that viruses must also have one. This normally consists of 7MeGpppN where N is normally adenine or guanine. The viruses encode a protein, normally a replicase, with a methyltransferase activity to allow this.

Some viruses are cap-snatchers. During this process, a ^{7m}G-capped host mRNA is recruited by the viral transcriptase complex and subsequently cleaved by a virally encoded endonuclease. The resulting capped leader RNA is used to prime transcription on the viral genome.

However some plant viruses do not use cap, yet translate efficiently due to cap-independent translation enhancers present in 5' and 3' untranslated regions of viral mRNA.

Readthrough

Some viruses (e.g. tobacco mosaic virus (TMV)) have RNA sequences that contain a "leaky" stop codon. In TMV 95% of the time the host ribosome will terminate the synthesis of the polypeptide at this codon but the rest of the time it continues past it. This means that 5% of the proteins produced are larger than and different from the others normally produced, which is a form of translational regulation. In TMV, this extra sequence of polypeptide is an RNA polymerase that replicates its genome.

Production of Sub-genomic RNAs

Some viruses use the production of subgenomic RNAs to ensure the translation of all proteins within their genomes. In this process the first protein encoded on the genome, and this the first to be translated, is a replicase. This protein will act on the rest of the genome producing negative strand sub-genomic RNAs then act upon these to form positive strand sub-genomic RNAs that are essentially mRNAs ready for translation.

Segmented Genomes

Some viral families, such as the *Bromoviridae* instead opt to have multipartite genomes, genomes split between multiple viral particles. For infection to occur, the plant must be infected with all particles across the genome. For instance *Brome mosaic virus* has a genome split between 3 viral particles, and all 3 particles with the different RNAs are required for infection to take place.

Polyprotein Processing

This strategy is adopted by viral genera such as the Potyviridae and Tymoviridae. The ribosome translates a single protein from the viral genome. Within the polyprotein is an enzyme (or enzymes) with proteinase function that is able to cleave the polyprotein into the various single proteins or just cleave away the protease, which can then cleave other polypeptides producing the mature proteins.

Well Understood Plant Viruses

Tobacco mosaic virus (TMV) and Cauliflower mosaic virus (CaMV) are frequently used in plant molecular biology. Of special interest is the CaMV 35S promoter, which is a very strong promoter most frequently used in plant transformations.

Pest Control

Pest control refers to the regulation or management of a species defined as a pest, and can be perceived to be detrimental to a person's health, the ecology or the economy. A practitioner of pest control is called an exterminator.

A crop duster applies low-insecticide bait that is targeted against western corn rootworms.

History

Pest control is at least as old as agriculture, as there has always been a need to keep crops free from pests. In order to maximize food production, it is advantageous to protect crops from competing species of plants, as well as from herbivores competing with humans.

The conventional approach was probably the first to be employed, since it is comparatively easy to destroy weeds by burning them or plowing them under, and to kill larger competing herbivores, such as crows and other birds eating seeds. Techniques such as crop rotation, companion planting (also known as intercropping or mixed cropping), and the selective breeding of pest-resistant cultivars have a long history.

In the UK, following concern about animal welfare, humane pest control and deterrence is gaining ground through the use of animal psychology rather than destruction. For instance, with the urban red fox which territorial behaviour is used against the animal, usually in conjunction with non-injurious chemical repellents. In rural areas of Britain, the use of firearms for pest control is quite common. Airguns are particularly popular for control of small pests such as rats, rabbits and grey squirrels, because of their lower power they can be used in more restrictive spaces such as gardens, where using a firearm would be unsafe.

Chemical pesticides date back 4,500 years, when the Sumerians used sulfur compounds as insecticides. The Rig Veda, which is about 4,000 years old, also mentions the use of poisonous plants for pest control. It was only with the industrialization and mechanization of agriculture in the 18th and 19th century, and the introduction of the insecticides pyrethrum and derris that chemical pest control became widespread. In the 20th century, the discovery of several synthetic insecticides, such as DDT, and herbicides boosted this development. Chemical pest control is still the predominant type of pest control today, although its long-term effects led to a renewed interest in traditional and biological pest control towards the end of the 20th century.

Causes

Many pests have only become a problem as a result of the direct actions by humans. Modifying these actions can often substantially reduce the pest problem. In the United States, raccoons caused a nuisance by tearing open refuse sacks. Many householders introduced bins with locking lids, which deterred the raccoons from visiting. House flies tend to accumulate wherever there is human activity and live in close association with people all over the world especially where food or food waste is exposed. Similarly, seagulls have become pests at many seaside resorts. Tourists would often feed the birds with scraps of fish and chips, and before long, the birds would rely on this food source and act aggressively towards humans.

Sign in Ilfracombe, England designed to help control seagull presence

Living organisms evolve and increase their resistance to biological, chemical, physical or any other form of control. Unless the target population is completely exterminated or is rendered incapable of reproduction, the surviving population will inevitably acquire a tolerance of whatever pressures are brought to bear - this results in an evolutionary arms race.

Types of Pest Control

Use of Pest-destroying Animals

Perhaps as far ago as 3000BC in Egypt, cats were being used to control pests of grain stores such as rodents. In 1939/40 a survey discovered that cats could keep a farm's population of rats down to a low level, but could not eliminate them completely. However, if the rats were cleared by trapping or poisoning, farm cats could stop them returning - at least from an area of 50 yards around a barn.

Ferrets were domesticated at least by 500 AD in Europe, being used as mousers. Mongooses have been introduced into homes to control rodents and snakes, probably at first by the ancient Egyptians.

Biological Pest Control

Biological pest control is the control of one through the control and management of natural predators and parasites. For example: mosquitoes are often controlled by putting *Bt Bacillus thuringiensis* ssp. *israelensis*, a bacterium that infects and kills mosquito larvae, in local water sources. The treatment has no known negative consequences on the remaining ecology and is safe for humans to drink. The point of biological pest control, or any natural pest control, is to eliminate a pest with minimal harm to the ecological balance of the environment in its present form.

Mechanical Pest Control

Mechanical pest control is the use of hands-on techniques as well as simple equipment and devices, that provides a protective barrier between plants and insects. For example: weeds can be controlled by being physically removed from the ground. This is referred to as tillage and is one of the oldest methods of weed control.

Physical Pest Control

Physical pest control is a method of getting rid of insects and small rodents by removing, attacking, setting up barriers that will prevent further destruction of one's plants, or forcing insect infestations to become visual.

Dog control van, Rekong Peo, Himachal Pradesh, India

Elimination of Breeding Grounds

Proper waste management and drainage of still water, eliminates the breeding ground of many pests.

Garbage provides food and shelter for many unwanted organisms, as well as an area where still water might collect and be used as a breeding ground by mosquitoes. Communities that have proper garbage collection and disposal, have far less of a problem with rats, cockroaches, mosquitoes, flies and other pests than those that don't.

Open air sewers are ample breeding ground for various pests as well. By building and maintaining a proper sewer system, this problem is eliminated.

Certain spectrums of LED light can "disrupt insects' breeding".

Poisoned Bait

Poisoned bait is a common method for controlling rat populations, however is not as effective when there are other food sources around, such as garbage. Poisoned meats have been used for centuries for killing off wolves, birds that were seen to threaten crops, and against other creatures. This can be a problem, since a carcass which has been poisoned will kill not only the targeted animal, but also every other animal which feeds on the carcass. Humans have also been killed by coming in contact with poisoned meat, or by eating an animal which had fed on a poisoned carcass. This tool is also used to manage several caterpillars e.g. Spodoptera litura, fruit flies, snails and slugs, crabs etc.

Field Burning

Traditionally, after a sugar cane harvest, the fields are all burned, to kill off any rodents, insects or eggs that might be in the fields.

Hunting

Historically, in some European countries, when stray dogs and cats became too numerous, local populations gathered together to round up all animals that did not appear to have an owner and kill them. In some nations, teams of rat-catchers work at chasing rats from the field, and killing them with dogs and simple hand tools. Some communities have in the past employed a bounty system, where a town clerk will pay a set fee for every rat head brought in as proof of a rat killing.

Traps

A variety of mouse traps and rat traps are available for mice and rats, including snap traps, glue traps and live catch traps.

Pesticides

Rodent bait station, Chennai, India

Spraying pesticides by planes, trucks or by hand is a common method of pest control. Crop dusters commonly fly over farmland and spray pesticides to kill off pests that would threaten the crops. However, some pesticides may cause cancer and other health problems, as well as harming wildlife.

Space Fumigation

A project that involves a structure be covered or sealed airtight followed by the introduction of a penetrating, deadly gas at a killing concentration a long period of time (24-72hrs.). Although expensive, space fumigation targets all life stages of pests.

Space Treatment

Residential & commercial building pest control service vehicle, Ypsilanti Township, Michigan

A long term project involving fogging or misting type applicators. Liquid insecticide is dispersed in the atmosphere within a structure. Treatments do not require the evacuation or airtight sealing of a building, allowing most work within the building to continue but at the cost of the penetrating effects. Contact insecticides are generally used, minimizing the long lasting residual effects. On August 10, 1973, the Federal Register printed the definition of Space treatment as defined by the U.S. Environmental Protection Agency (EPA):

> “the dispersal of insecticides into the air by foggers, misters, aerosol devices or vapor dispensers for control of flying insects and exposed crawling insects”

Sterilization

Laboratory studies conducted with U-5897 (3-chloro-1,2-propanediol) were attempted in the early 1970s although these proved unsuccessful. Research into sterilization bait is ongoing.

In 2013, New York City tested sterilization traps in a $1.1 million study. The result was a 43% reduction in rat populations. The Chicago Transit Authority plans to test sterilization control in spring 2015. The sterilization method doesn't poison the rats or humans.

Destruction of Infected Plants

Forest services sometimes destroy all the trees in an area where some are infected with insects, if seen as necessary to prevent the insect species from spreading. Farms infested with certain insects, have been burned entirely, to prevent the pest from spreading elsewhere.

Natural Rodent Control

Several wildlife rehabilitation organizations encourage natural form of rodent control through exclusion and predator support and preventing secondary poisoning altogether.

Example of House mouse infestation

he United States Environmental Protection Agency agrees, noting in its Proposed Risk Mitigation Decision for Nine Rodenticides that "without habitat modification to make areas less attractive to commensal rodents, even eradication will not prevent new populations from recolonizing the habitat."

Repellents

- Balsam fir oil from the tree *Abies balsamea* is an EPA approved non-toxic rodent repellent.
- *Acacia polyacantha* subsp. *campylacantha* root emits chemical compounds that repel animals including crocodiles, snakes and rats.

Plant Disease Epidemiology

Plant disease epidemiology is the study of disease in plant populations. Much like diseases of humans and other animals, plant diseases occur due to pathogens such as bacteria, viruses, fungi, oomycetes, nematodes, phytoplasmas, protozoa, and parasitic plants. Plant disease epidemiologists strive for an understanding of the cause and effects of disease and develop strategies to intervene in situations where crop losses may occur. Typically successful intervention will lead to a low enough level of disease to be acceptable, depending upon the value of the crop.

Plant disease epidemiology is often looked at from a multi-disciplinary approach, requiring biological, statistical, agronomic and ecological perspectives. Biology is necessary for understanding the pathogen and its life cycle. It is also necessary for understanding the physiology of the crop and how the pathogen is adversely affecting it. Agronomic practices often influence disease incidence for better or for worse. Ecological influences are numerous. Native species of plants may serve as reservoirs for pathogens that cause disease in crops. Statistical models are often applied in order to summarize and describe the complexity of plant disease epidemiology, so that disease processes can be more readily understood. For example, comparisons between patterns of disease progress for different diseases, cultivars, management strategies, or environmental settings can help in determining how plant diseases may best be managed. Policy can be influential in the occurrence of diseases, through actions such as restrictions on imports from sources where a disease occurs.

In 1963 J. E. van der Plank published "Plant Diseases: Epidemics and Control", a seminal work that created a theoretical framework for the study of the epidemiology of plant diseases. This book provides a theoretical framework based on experiments in many different host pathogen systems and moved the study of plant disease epidemiology forward rapidly, especially for fungal foliar pathogens. Using this framework we can now model and determine thresholds for epidemics that take place in a homogeneous environment such as a mono-cultural crop field.

Elements of an Epidemic

Disease epidemics in plants can cause huge losses in yield of crops as well threatening to wipe out an entire species such as was the case with Dutch Elm Disease and could occur with Sudden Oak Death. An epidemic of potato late blight, caused by *Phytophthora infestans*, led to the Great Irish Famine and the loss of many lives.

Commonly the elements of an epidemic are referred to as the "disease triangle": a susceptible host, pathogen, and conducive environment. For disease to occur all three of these must be present. Below is an illustration of this point. Where all three items meet there is disease. The fourth element missing from this illustration for an epidemic to occur, is time. As long as all three of these elements are present disease can initiate, an epidemic will only ensue if all three continue to be present. Any one of the three might be removed from the equation though. The host might out-grow susceptibility as with high-temperature adult-plant resistance, the environment changes and is not conducive for the pathogen to cause disease, or the pathogen is controlled through a fungicide application for instance.

Sometimes a fourth factor of time is added as the time at which a particular infection occurs, and the length of time conditions remain viable for that infection, can also play an important role in epidemics. The age of the plant species can also play a role, as certain species change in their levels of disease resistance as they mature; a process known as ontogenic resistance.

If all of the criteria are not met, such as a susceptible host and pathogen are present but the environment is not conducive to the pathogen infecting and causing disease, disease cannot occur. For example, corn is planted into a field with corn residue that has the fungus *Cercospora zea-maydis*, the causal agent of Grey leaf spot of corn, but if the weather is too dry and there is no leaf wetness the spores of the fungus in the residue cannot germinate and initiate infection.

Likewise, it stands to reason if the host is susceptible and the environment favours the development of disease but the pathogen is not present there is no disease. Taking the example above, the corn is planted into a ploughed field where there is no corn residue with the fungus *Cercospora zea-maydis*, the causal agent of Grey leaf spot of corn, present but the weather means long periods of leaf wetness, there is no infection initiated.

When a pathogen requires a vector to be spread then for an epidemic to occur the vector must be plentiful and active.

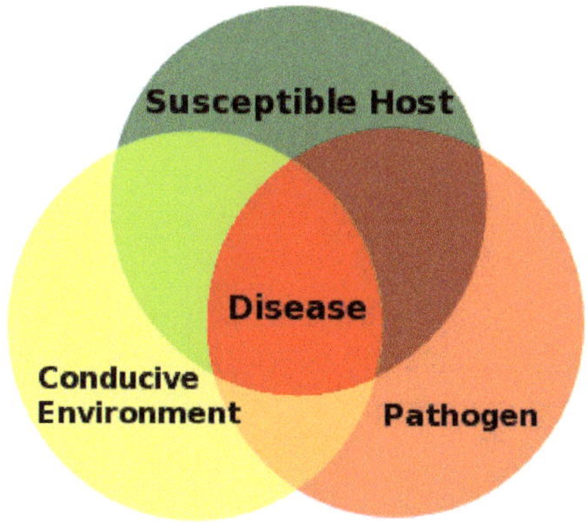

Plant disease triangle illustration

Types of Epidemics

Monocyclic epidemics are caused by pathogens with a low birth rate and death rate, meaning they only have one infection cycle per season. They are typical of soil-borne diseases such as Fusarium wilt of flax. Polycyclic epidemics are caused by pathogens capable of several infection cycles a season. They are most often caused by airborne diseases such as powdery mildew. Bimodal polycyclic epidemics can also occur. For example, in brown rot of stone fruits the blossoms and the fruits are infected at different times.

For some diseases it is important to consider the disease occurrence over several growing seasons, especially if growing the crops in monoculture year after year or growing perennial plants. Such conditions can mean that the inoculum produced in one season can be carried over to the next leading to a build of an inoculum over the years. In the tropics there are no clear-cut breaks between growing seasons as there are in temperate regions and this can lead to accumulation of inoculum.

Epidemics that occur under these conditions are referred to as *polyetic* epidemics and can be caused by both monocyclic and polycyclic pathogens. Apple powdery mildew is an example of a polyetic epidemic caused by a polycyclic pathogen and Dutch Elm disease a polyetic epidemic caused by a monocyclic pathogen.

Employment

Plant disease epidemiologists are typically employed as researchers by universities, or governmental institutions such as the USDA. However, private companies in agricultural fields also employ epidemiologists.

Morphological Symptoms of Plant Diseases

Thousands of plant diseases have been recorded throughout the world, many of these causing heavy crop loses. Early detection and accurate diagnosis is essential for the effective management of plant disease. Thus the first step in studying any disease is its timely detection of the diseased plant. Quick initial detection is largely based on the signs and symptoms of disease.

Signs are the visible physical presence of either the pathogen itself or the structures formed by the pathogen. Common examples of easily detected signs are those such as the fungal mycelia and spore masses of downy mildews observed on infected leaves and the bacterial ooze of *Xanthomonas* leaf streak disease on rice.

Symptoms are the visible changes that occur in the host plant in response to infection by pathogens. For any disease in a given plant, there is the characteristic expression of symptoms, usually occurring in a sequential series during the course of the disease. This series of symptoms depicting the disease picture is referred to as the disease syndrome.

Morphological symptoms may be exhibited by the entire plant or by any organ of the plant. These have been categorized into different groups for easy of study. Primarily, morphological symptoms of plant diseases can be categorized into 6 different types.

- Necroses
- Growth abnormalities
- Metaplastic symptoms
- Proleptic symptoms
- Color changes
- Wilts

Necroses

Necroses are caused due to necrosis or death of plant cells. The affected plant tissue usually turns brown to black in color. Necrotic symptoms could appear in any part of the plant such as in storage organs, in green tissues, or in woody tissues.

Necrosis in Storage Organs

Death of cells in storage organs terminates in decomposition or decay referred to as a rot. Two types of rots are identified as Dry rot and Wet rot on storage tissues.

Soft rots are those where the pathogen breaks down the host cell walls, resulting in the exudation of juices from the infected tissue. The organ becomes mushy or pulpy and a foul smell often develops due to colonization by secondary invaders. Many fungi and bacteria cause soft rots on several fruits and vegetables. Species of the fungus, *Rhizopus* and bacterium *Erwinia* are two such commonly found pathogens causing soft rots.

In a dry rot, the storage organ becomes hard and dry, and in some diseases, there is rapid loss of water and the infected organs become shriveled, wrinkled, and leathery. Dry rots showing such symptoms are referred to as mummifications.

Necrosis in Green Tissues

Necroses on green tissue are termed differently based on the nature of symptoms and the type of green tissue. The term, damping off refers to the sudden wilting and topping over of seedlings as a result of extensive necrosis of tender tissue of the roots and stem near the soil line, due to the attack of soil-borne pathogens such as fungus, *Pythium*. This fungus is known to cause damping off in an assortment of seedlings such as that of brinjal, chilli, mung beans, tobacco, tomato, and *Cucurbita*.

A spot refers to a well-defined area of gray or brown necrotic tissue. Spots are very common on leaves and fruits and are probably the most familiar necrotic symptom. Sometimes the necrotic tissue within a leaf spot may crack and fall off from the surrounding green tissue leaving an empty space. Such a symptom is known as a shot hole.

Minute or very small spots are sometimes referred to as flecks or specks.

When dark mycelia of a fungal pathogen appear on the surface of necrotic spot, blotting the leaves, shoots, an stems as large and irregular spots, the symptom is referred to as a blotch.

Both streaks and stripes occur in grasses and are elongated areas having dead cells. Streaks occur along the stem and veins, while stripes are in the laminar tissues between veins.

Net necrosis is a symptom resulting from an irregular pattern of anastomoses between streaks or stripes.

Blights are characterized by the rapid death of entire leaves including the veins or parts of the leaves. Blights also could occur on flowers and stems.

Scorches resemble blights, but there necrosis occurs in irregular patterns between veins and along leaf margins.

Firing is sudden drying, collapse and death of entire leaves. Firing occurs in response to the activity of root rot and vascular wilt pathogens.

Scald is the term used to describe the blanching of epidermal and adjacent tissues of fruits and occasionally of leaves.

The sudden death of unopened buds or inflorescence is referred to as blast.

Extensive necrosis of fruits that resemble in premature dropping is called shelling.

Necrosis in Woody Tissues

Necrosis of woody tissue often brings about various types of die-back symptoms. Dieback is the extensive necrosis of a shoot from its tip downwards.

Restricted necrosis of the bark and cortical tissue of stems and roots is termed as a canker. In cankers, necrotic tissue in the sunken lesions is sharply limited, usually by a callus from adjacent healthy tissue.

When woody tissues are diseased, they may exude different kinds of substances. When the exudate is gummy, the symptom is called gummosis, while it is resinosis when it is resinous. If the exudate is neither gummy nor resinous, it is described as bleeding.

Abnormalities in Growth

Many disease symptoms are associated with growth changes in diseased plants. These could be caused by either reduced growth due to hypoplasia and atrophy or excessive growth due to hyperplasia and hypertrophy.

Hypoplasia & Atrophy

Hypoplasia is the failure of plants or plant organs to develop fully due to a decreased production of the number of cells. Hypoplasia results in plants or plant parts of sub-normal size.

Atrophy is the reduction in the size of plant cells produced. This also results in stunted plants or plant parts.

Dwarfing is the failure of a plant or a plant part to attain its full size.

Rosetting is a condition where the internode of a plant do not elongate, and hence, the leaves appear close together in a cluster.

Hyperplasia & Hypertrophy

Hyperplasia is the enlargement of a plant tissue due to excessive increase in the number of plant cells produced. Hyperplasia results in overdevelopment in size of plants or plant organs.

Hypertrophy is excessive growth due to the enlargement of individual cells. This condition also results in the overdevelopment in size of plants or plant organs.

Hyperplasia and hypertrophy could result in the enlargement of leaves and fruits, and the enlargement of stems and roots.

Enlargement of Leaves and Fruits

Several symptoms expressing enlargement of leaves and fruits are commonly observed among diseased plants.

Curling which is the bending of the shoot or the rolling of the leaf, is a result of over-growth on one side of an organ. Often viral diseases cause such leaf distortions due to irregular growth of the lamina. Extreme reduction of the leaf lamina brings about the symptom known as the Shoe-string effect.

The puckering or crinkling of leaves due to different growth rates in adjacent tissue is known as savoying.

Overgrowth of epidermal and underlying tissues of leaves, stems, fruits and tubers may result scab formation. Scab consists of raised, rough, and discrete lesions. These are often sunken and cracked, giving a typical scabby appearance.

Localized swellings or enlargement of epidermal cells due to excessive accumulation of water is termed intermuscence and the diagnostic symptom is the appearance of a blister.

Enlargement of Stems and Roots

Symptoms causing enlargement of stems and roots are termed differently based on their nature. Excessive accumulation of food material in stems, above a constricted area produces a swelling termed sarcody.

Localized swellings that involve entire organs are termed tumefaction. Commonly exhibited tumefactions are galls, clubs, and knots.

Excessive development of adventitious organs results in fasciculation, that is the clustering of organs around a focal point. Such examples include witch's broom and hairy root. Witch's broom is a broom-like mass proliferation due to the dense clustering of branches of woody plants while hairy root results due to excessive development of roots.

Fasciation is the broadening or flattening of cylindrical organs such as stems. The continued development of any organ after it has reached a stage beyond which it normally does not grow is known as proliferation.

The outgrowth of tissue in response to wounding is known as a callus. Callus formation is found to form around most cankers.

Metaplastic Symptoms

Metaplastic symptoms are those which form when tissues change from one form to another. Such symptoms include phyllody, the development of floral organs into leaf-like structures, juvenillody, the development of juvenile seedlings on mature plants and russeting, a superficial browning of surfaces of fruits and tubers due to suberization.

Proleptic Symptoms

Proleptic symptoms result from the development of tissues earlier than usual. Examples include prolepsis, the premature development of a shoot from a bud, proleptic abscission, the premature formation of abscission layers and restoration, the unexpected development of organs that are normally rudimentary.

Color Changes

Changes in the color of plant tissue are a common symptom of plant disease. Often these color changes are brought about by the yellowing of normal green tissue due to the destruction of chlorophyll or a failure to form chlorophyll. Such repression of leaf color may be complete or partial.

When color repression is complete, it is known as albication. However, the more common, partial repression is referred to as chlorosis.

Patches of green tissue alternating with chlorotic areas are described as a mosaic. Mosaic is a symptom caused by many viruses. Based on the intensity and the pattern of discoloration, mosaics are termed differently. Irregular patches of distinct light and dark areas are known as mottling. Streaking and ring spots are still other distinct types of discolorations. Ring spots are circular masses of chlorosis with a green center. Vein clearing and vein banding are yet other common color changes on leaves.

Chlorophyll may also develop in tissues normally devoid of it. Thus usually white or colored tissue becomes green in color. This is called as virescence.

Anthocyanescence is due to the overdevelopment of anthocyanin and result in the development of a purplish coloration. Color changes can also take place in flowers. Such an example is the color break virus-affected tulips.

Wilts

Wilting is due to loss of turgor in plant tissue resulting in the dropping of plant parts. They are common symptom in diseases where the pathogen or the toxic metabolites it produces affects the vascular tissue of the host plant. Interference in water transport brought about by the infection of these vascular pathogens leads to wilting. Unlike wilting due to low soil moisture, wilting due to the activity of these pathogens cannot be overcome by watering the plants. Infected plants eventually die.

Plant Disease Resistance

Cankers caused Chestnut blight a disease that affects the chestnut tree

Plant disease resistance protects plants from pathogens in two ways: mechanisms and by infection-induced responses of the immune system. Relative to a susceptible plant, disease resistance is the reduction of pathogen growth on or in the plant, while the term disease tolerance describes plants that exhibit little disease damage despite substantial pathogen levels. Disease outcome is determined by the three-way interaction of the pathogen, the plant and the environmental conditions (an interaction known as the disease triangle).

Defense-activating compounds can move cell-to-cell and systemically through the plant vascular system. However, plants do not have circulating immune cells, so most cell types exhibit a broad suite of antimicrobial defenses. Although obvious *qualitative* differences in disease resistance can be observed when multiple specimens are compared (allowing classification as "resistant" or "susceptible" after infection by the same pathogen strain at similar inoculum levels in similar environments), a gradation of *quantitative* differences in disease resistance is more typically observed between plant strains or genotypes. Plants consistently resist certain pathogens but succumb to others; resistance is usually pathogen species- or pathogen strain-specific.

Background

Plant disease resistance is crucial to the reliable production of food, and it provides significant reductions in agricultural use of land, water, fuel and other inputs. Plants in both natural and cultivated populations carry inherent disease resistance, but this has not always protected them.

The late blight Irish potato famine of the 1840s was caused by the oomycete Phytophthora infestans. The world's first mass-cultivated banana cultivar Gros Michel was lost in the 1920s to Panama disease caused by the fungus Fusarium oxysporum. The current wheat stem, leaf, and yellow stripe rust epidemics spreading from East Africa into the Indian subcontinent are caused by rust fungi Puccinia graminis and P. striiformis. Other epidemics include Chestnut blight, as well as recurrent severe plant diseases such as Rice blast, Soybean cyst nematode, Citrus canker.

Plant pathogens can spread rapidly over great distances, vectored by water, wind, insects, and humans. Across large regions and many crop species, it is estimated that diseases typically reduce plant yields by 10% every year in more developed nations or agricultural systems, but yield loss to diseases often exceeds 20% in less developed settings, an estimated 15% of global crop production.

However, disease control is reasonably successful for most crops. Disease control is achieved by use of plants that have been bred for good resistance to many diseases, and by plant cultivation approaches such as crop rotation, pathogen-free seed, appropriate planting date and plant density, control of field moisture and pesticide use.

Viral Disease Common Mechanisms

Pre-formed Structures and Compounds

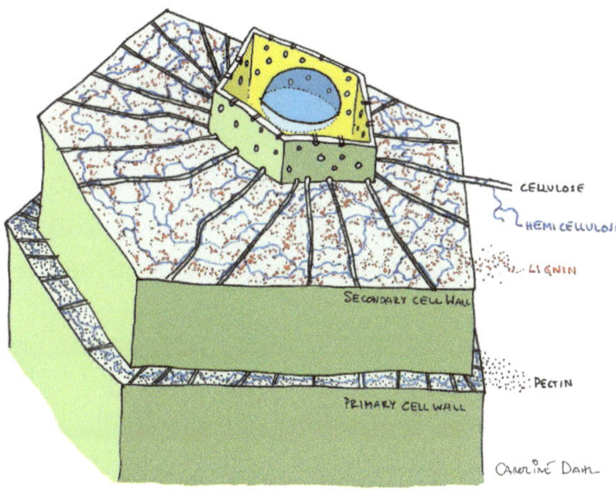

secondary plant wall

- Plant cuticle/surface
- Plant cell walls
- Antimicrobial chemicals (for example: glucosides, saponins)

- Antimicrobial proteins
- Enzyme inhibitors
- Detoxifying enzymes that break down pathogen-derived toxins
- Receptors that perceive pathogen presence and activate inducible plant defences

Inducible Post-infection Plant Defenses

- Cell wall reinforcement (callose, lignin, suberin, cell wall proteins)
- Antimicrobial chemicals, including reactive oxygen species such as hydrogen peroxide or peroxynitrite, or more complex phytoalexins such as genistein or camalexin
- Antimicrobial proteins such as defensins, thionins, or PR-1
- Antimicrobial enzymes such as chitinases, beta-glucanases, or peroxidases
- Hypersensitive response - a rapid host cell death response associated with defence mediated by "Resistance genes."(Bryant, Tracy 2008).

Variable Resistance

Even in susceptible plants to obligate parasites the tissue resistance changes due to ontogeny and to influence of external conditions. This resistance can be measured by the value of redox potential of electron carriers, which is produced in the plant by enzymatic reactions associated with respiration. Electron carriers are water-soluble and are not oxidized by air oxygen. There is not free oxygen in the cells, all the oxidizations and reductions take place enzymatically. These reactions are highly specific for the plant species. The host and parasite have different electron carriers.

Immune System

The plant immune system consists of two interconnected tiers of receptors, one outside and one inside the cell. Both systems sense the intruder, respond to the intrusion and optionally signal to the rest of the plant and sometimes to neighboring plants that the intruder is present. The two systems detect different types of pathogen molecules and classes of plant receptor proteins

The first tier is primarily governed by pattern recognition receptors that are activated by recognition of evolutionarily conserved pathogen or microbial–associated molecular patterns (PAMPs or MAMPs, here P/MAMP). Activation of PRRs leads to intracellular signaling, transcriptional reprogramming, and biosynthesis of a complex output response that limits colonization. The system is known as PAMP-Triggered Immunity (PTI)"(JonesDangl2010).

The second tier (again, primarily), effector-triggered immunity (ETI), consists of another set of receptors, (nucleotide-binding)They operate within the cell, encoded by R genes. The presence of specific pathogen "effectors" activates specific NLR proteins that limit pathogen proliferation.

Receptor responses include ion channel gating, oxidative burst, cellular redox changes, or protein kinase cascades that directly activate cellular changes (such as cell wall reinforcement or antimicrobial production), or activate changes in gene expression that then elevate other defensive responses

Plant immune systems show some mechanistic similarities with the immune systems of insects and mammals, but also exhibit many plant-specific characteristics. Plants can sense the presence of pathogens and the effects of infection via activated by touch]. Rice Universi

PAMP-triggered Immunity

PAMP-Triggered Immunity conserved molecules that inhabit multiple pathogen genera are classified as MAMPs by some researchers. The defenses induced by MAMP perception are sufficient to repel most pathogens. However, pathogen effector proteins are adapted to suppress basal defenses such as PTI

Effector Triggered Immunity

Effector Triggered Immunity (ETI) is activated by the presence of pathogen effectors. The ETI immune response is reliant on R genes, and is activated by specific pathogen strains. As with PTI, many specific examples of apparent ETI violate commoMost plant immune systems carry a repertoire of 100-600 different R genes that mediate resistance to various virus .Plant ETI often cause an apoptotic hypersensitive response.(Odds Rathjen2010).

R Genes and R Proteins

Plants have evolved R genes (resistance genes) whose products allow recognition of specific pathogen effectors, either through direct binding or by recognition of the effector's alteration of a host protein. These virulence factors drove co-evolution of plant resistant genes to combat the pathogens' Avr (avirulent) genes. Many R genes encode NB-LRR proteins (nucleotide-binding/leucine-rich repeat domains, also known as NLR proteins or STAND proteins, among other names).

R gene products control a broad set of disease resistance responses whose induction is often sufficient to stop further pathogen growth/spread. Each plant genome contains a few hundred apparent R genes. Studied R genes usually confer specificity for particular pathogen strains. As first noted by Harold Flor in his mid-20th century formulation of the gene-for-gene relationship, the plant R gene and the pathogen Avr gene must have matched specificity for that R gene to confer resistance, suggesting a receptor/ligand interaction for Avr and R genes. Alternatively, an effector can modify its host cellular target (or a molecular decoy of that target) activating an NLR associated with the

Effector Biology

So-called "core" effectors are defined operationally by their wide distribution across the population of a particular pathogen and their substantial contribution to pathogen virulence. Genomics can be used to identify core effectors, which can then functionally define new R alleles, which can serve as breeding targets.

RNA Silencing and Systemic Acquired Resistance Elicited by Prior Infections

Against viruses, plants often induce pathogen-specific gene silencing mechanisms mediated by RNA interference. T

Plant immune systems also can respond to an initial infection in one part of the plant by physiologically elevating the capacity for a successful defense response in other parts. Such responses include systemic acquired resistance, largely mediated by salicylic acid-dependent pathways, and induced systemic resistance, largely mediated.

Species-level Resistance

In a small number of cases, plant genes are effective against an entire pathogen species, even though that species that is pathogenic on other genotypes of that host species. Examples include barley MLO against powdery mildew, wheat Lr34 against leaf rust and wheat Yr36 against stripe rust. An array of mechanisms for this type of resistance may exist depending on the particular gene and plant-pathogen combination. Other reasons for effective plant immunity can include a lack of coadaptation (the pathogen and/or plant lack multiple mechanisms needed for colonization and growth within that host species), or a particularly effective suite of pre-formed defenses.

Signaling Mechanisms

Perception of Pathogen Presence

Plant defense signaling is activated by pathogen-detecting receptors. The activated receptors frequently elicit reactive oxygen and nitric oxide production, calcium, potassium and proton ion fluxes, altered levels of salicylic acid and other hormones and activation of MAP kinases and other specific protein kinases. These events in turn typically lead to the modification of proteins that control gene transcription, and the activation of defense-associated gene expression.

In addition to PTI and ETI, plant defenses can be activated by the sensing of damage-associated compounds (DAMP), such as portions of the plant cell wall released during pathogenic infection. Many receptors for MAMPs, effectors and DAMPs have been discovered. Effectors are often detected by NLRs, while MAMPs and DAMPs are often detected by transmembrane receptor-kinases that carry LRR or LysM extracellular domains.

Transcription Factors and the Hormone Response

Numerous genes and/or proteins as well as other molecules have been identified that mediate plant defense signal transduction. Cytoskeleton and vesicle trafficking dynamics help to orient plant defense responses toward the point of pathogen attack.

Mechanisms of Transcription Factors and Hormones

Plant immune system activity is regulated in part by signaling hormones such as:

- Salicylic acid
- Jasmonic acid
- Ethylene

There can be substantial cross-talk among these pathways.

Regulation by Degradation

As with many signal transduction pathways, plant gene expression during immune responses can be regulated by degradation. This often occurs when hormone binding to hormone receptors stimulates ubiquitin-associated degradation of repressor proteins that block expression of certain genes. The net result is hormone-activated gene expression. Examples:

- Auxin: binds to receptors that then recruit and degrade repressors of transcriptional activators that stimulate auxin-specific gene expression.
- Jasmonic acid: similar to auxin, except with jasmonate receptors impacting jasmonate-response signaling mediators such as JAZ proteins.
- Gibberellic acid: Gibberellin causes receptor conformational changes and binding and degradation of Della proteins.
- Ethylene: Inhibitory phosphorylation of the EIN2 ethylene response activator is blocked by ethylene binding. When this phosphorylation is reduced, EIN2 protein is cleaved and a portion of the protein moves to the nucleus to activate ethylene-response gene expression.

Ubiquitin and E3 Signaling

Ubiquitination plays a central role in cell signaling that regulates processes including protein degradation and immunological response. Although one of the main functions of ubiquitin is to target proteins for destruction, it is also useful in signaling pathways, hormone release, apoptosis and translocation of materials throughout the cell. Ubiquitination is a component of several immune responses. Without ubiquitin's proper functioning, the invasion of pathogens and other harmful molecules would increase dramatically due to weakened immune defenses.

E3 Signaling

The E3 Ubiquitin ligase enzyme is a main component that provides specificity in protein degradation pathways, including immune signaling pathways. The E3 enzyme components can be grouped by which domains they contain and include several types. These include the Ring and U-box single subunit, HECT, and CRLs. Plant signaling pathways including immune responses are controlled by several feedback pathways, which often include negative feedback; and they can be regulated by De-ubiquitination enzymes, degradation of transcription factors and the degradation of negative regulators of transcription.

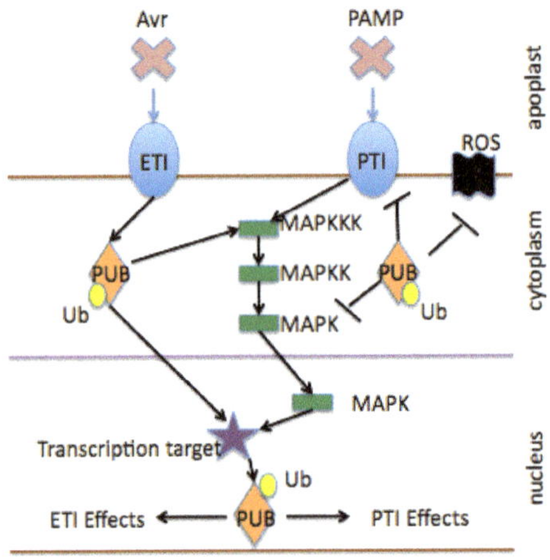

This image depicts the pathways taken during responses in plant immunity. It highlights the role and effect ubiquitin has in regulating the pathway.

Plant Breeding for Disease Resistance

Plant breeders emphasize selection and development of disease-resistant plant lines. Plant diseases can also be partially controlled by use of pesticides and by cultivation practices such as crop rotation, tillage, planting density, disease-free seeds and cleaning of equipment, but plant varieties with inherent (genetically determined) disease resistance are generally preferred. Breeding for disease resistance began when plants were first domesticated. Breeding efforts continue because pathogen populations are under selection pressure for increased virulence, new pathogens appear, evolving cultivation practices and changing climate can reduce resistance and/or strengthen pathogens, and plant breeding for other traits can disrupt prior resistance. A plant line with acceptable resistance against one pathogen may lack resistance against others.

Breeding for resistance typically includes:

- Identification of plants that may be less desirable in other ways, but which carry a useful disease resistance trait, including wild strains that often express enhanced resistance.
- Crossing of a desirable but disease-susceptible variety to another variety that is a source of resistance.
- Growth of breeding candidates in a disease-conducive setting, possibly including pathogen inoculation. Attention must be paid to the specific pathogen isolates, to address variability within a single pathogen species.
- Selection of disease-resistant individuals that retain other desirable traits such as yield, quality and including other disease resistance traits.

Resistance is termed *durable* if it continues to be effective over multiple years of widespread use as pathogen populations evolve. "Vertical resistance" is specific to certain races or strains of a pathogen species, is often controlled by single R genes and can be less durable. Hoizontal or broad-spectrum resistance against an entire pathogen species is often only incompletely effective, but more durable, and is often controlled by many genes that segregate in breeding populations.

Crops such as potato, apple, banana and sugarcane are often propagated by vegetative reproduction to preserve highly desirable plant varieties, because for these species, outcrossing seriously disrupts the preferred traits. Vegetatively propagated crops may be among the best targets for resistance improvement by the biotechnology method of plant transfor-mation to manage genes that affect disease resistance.

Scientific breeding for disease resistance originated with Sir Rowland Biffen, who identified a single recessive gene for resistance to wheat yellow rust. Nearly every crop was then bred to include disease resistance (R) genes, many by introgression from compatible wild relatives.

GM or Transgenic Engineered Disease Resistance

The term GM ("genetically modified") is often used as a synonym of transgenic to refer to plants modified using recombinant DNA technologies. Plants with transgenic/GM disease resistance against insect pests have been extremely successful as commercial products, especially in maize and cotton, and are planted annually on over 20 million hectares in over 20 countries worldwide. Transgenic plant disease resistance against microbial patho-gens was first demonstrated in 1986. Expression of viral coat protein gene sequences conferred virus resistance via small RNAs. This proved to be a widely applicable mechanism for inhibiting viral replication. Combining coat protein genes from three different viruses, scientists developed squash hybrids with field-validated, multiviral resistance. Similar levels of resistance to this vari-ety of viruses had not been achieved by conventional breeding.

A similar strategy was deployed to combat papaya ringspot virus, which by 1994 threatened to destroy Hawaii's papaya industry. Field trials demonstrated excellent efficacy and high fruit quality. By 1998 the first transgenic virus-resistant papaya was approved for sale. Disease resistance has been durable for over 15 years. Transgenic papaya accounts for ~85% of Hawaiian production. The fruit is approved for sale in the U.S., Canada and Japan.

Potato lines expressing viral replicase sequences that confer resistance to potato leafroll virus were sold under the trade names NewLeaf Y and NewLeaf Plus, and were widely accepted in commercial production in 1999-2001, until McDonald's Corp. decided not to purchase GM potatoes and Monsanto decided to close their NatureMark potato business. NewLeaf Y and NewLeaf Plus potatoes carried two GM traits, as they also expressed Bt-mediated resistance to Colorado potato beetle.

No other crop with engineered disease resistance against microbial pathogens had reached the market by 2013, although more than a dozen were in some state of development and testing.

Examples of transgenic disease resistance projects

Publication year	Crop	Disease resistance	Mechanism	Development status
2012	Tomato	Bacterial spot	R gene from pepper	8 years of field trials
2012	Rice	Bacterial blight and bacterial streak	Engineered E gene	Laboratory
2012	Wheat	Powdery mildew	Overexpressed R gene from wheat	2 years of field trials at time of publication
2011	Apple	Apple scab fungus	Thionin gene from barley	4 years of field trials at time of publication

2011	Potato	Potato virus Y	Pathogen-derived resistance	1 year of field trial at time of publication
2010	Apple	Fire blight	Antibacterial protein from moth	12 years of field trials at time of publication
2010	Tomato	Multibacterial resistance	PRR from *Arabidopsis*	Laboratory scale
2010	Banana	Xanthomonas wilt	Novel gene from pepper	Now in field trial
2009	Potato	Late blight	R genes from wild relatives	3 years of field trials
2009	Potato	Late blight	R gene from wild relative	2 years of field trials at time of publication
2008	Potato	Late blight	R gene from wild relative	2 years of field trials at time of publication
2008	Plum	Plum pox virus	Pathogen-derived resistance	Regulatory approvals, no commercial sales
2005	Rice	Bacterial streak	R gene from maize	Laboratory
2002	Barley	Stem rust	Resting lymphocyte kinase (RLK) gene from resistant barley cultivar	Laboratory
1997	Papaya	Ring spot virus	Pathogen-derived resistance	Approved and commercially sold since 1998, sold into Japan since 2012
1995	Squash	Three mosaic viruses	Pathogen-derived resistance	Approved and commercially sold since 1994
1993	Potato	Potato virus X	Mammalian interferon-induced enzyme	3 years of field trials at time of publication

PRR Transfer

Research aimed at engineered resistance follows multiple strategies. One is to transfer useful PRRs into species that lack them. Identification of functional PRRs and their transfer to a recipient species that lacks an orthologous receptor could provide a general pathway to additional broadened PRR repertoires. For example, the *Arabidopsis* PRR *EF-Tu* receptor (EFR) recognizes the bacterial translation elongation factor *EF-Tu*. Research performed at Sainsbury Laboratory demonstrated that deployment of EFR into either *Nicotiana benthamiana*or *Solanum lycopersicum* (tomato), which cannot recognize *EF-Tu*, conferred resistance to a wide range of bacterial pathogens. EFR expression in tomato was especially effective against the widespread and devastating soil bacterium Ralstonia solanacearum. Conversely, the tomato PRR *Verticillium 1* (*Ve1*) gene can be transferred from tomato to *Arabidopsis*, where it confers resistance to race 1 Verticillium isolates.

Stacking

The second strategy attempts to deploy multiple NLR genes simultaneously, a breeding strategy known as stacking. Cultivars generated by either DNA-assisted molecular breeding or gene transfer will likely display more durable resistance, because pathogens would have to mutate multiple effector genes. DNA sequencing allows researchers to functionally "mine" NLR genes from multiple species/strains.

The *avrBs2* effector gene from *Xanthomona perforans* is the causal agent of bacterial spot disease of pepper and tomato. The first "effector-rationalized" search for a potentially durable R gene followed the finding that *avrBs2* is found in most disease-causing *Xanthomonas* species and is required for pathogen fitness. The *Bs2* NLR gene from the wild pepper, *Capsicum chacoense*, was moved into tomato, where it inhibited pathogen growth. Field trials demonstrated robust resistance without bactericidal chemicals. However, rare strains of *Xanthomonas* overcame *Bs2*-mediated resistance in pepper by acquisition of *avrBs2* mutations that avoid recognition but retain virulence. Stacking R genes that each recognize a different core effector could delay or prevent adaptation.

More than 50 loci in wheat strains confer disease resistance against wheat stem, leaf and yellow stripe rust pathogens. The Stem rust 35 (*Sr35*) NLR gene, cloned from a diploid relative of cultivated wheat, *Triticum monococcum*, provides resistance to wheat rust isolate *Ug99*. Similarly, *Sr33*, from the wheat relative *Aegilops tauschii*, encodes a wheat ortholog to barley *Mla* powdery mildew–resistance genes. Both genes are unusual in wheat and its relatives. Combined with the *Sr2* gene that acts additively with at least Sr33, they could provide durable disease resistance to *Ug99* and its derivatives.

Executor Genes

Another class of plant disease resistance genes opens a "trap door" that quickly kills invaded cells, stopping pathogen proliferation. Xanthomonas and Ralstonia transcription activator–like (TAL) effectors are DNA-binding proteins that activate host gene expression to enhance pathogen virulence. Both the rice and pepper lineages independently evolved TAL-effector binding sites that instead act as an executioner that induces hypersensitive host cell death when up-regulated. *Xa27* from rice and Bs3 and Bs4c from pepper, are such "executor" (or "executioner") genes that encode non-homologous plant proteins of unknown function. Executor genes are expressed only in the presence of a specific TAL effector.

Engineered executor genes were demonstrated by successfully redesigning the pepper *Bs3* promoter to contain two additional binding sites for TAL effectors from disparate pathogen strains. Subsequently, an engineered executor gene was deployed in rice by adding five TAL effector binding sites to the *Xa27* promoter. The synthetic *Xa27* construct conferred resistance against Xanthomonas bacterial blight and bacterial leaf streak species.

Host Susceptibility Alleles

Most plant pathogens reprogram host gene expression patterns to directly benefit the pathogen. Reprogrammed genes required for pathogen survival and proliferation can be thought of as "disease-susceptibility genes." Recessive resistance genes are disease-susceptibility candidates. For example, a mutation disabled an *Arabidopsis* gene encoding pectate lyase (involved in cell wall degradation), conferring resistance to the powdery mildew pathogen *Golovinomyces cichoracearum*. Similarly, the Barley *MLO* gene and spontaneously mutated pea and tomato *MLO* orthologs also confer powdery mildew resistance.

Lr34 is a gene that provides partial resistance to leaf and yellow rusts and powdery mildew in wheat. *Lr34* encodes an adenosine triphosphate (ATP)–binding cassette (ABC) transporter. The dominant allele that provides disease resistance was recently found in cultivated wheat (not in wild strains) and, like *MLO* provides broad-spectrum resistance in barley.

Natural alleles of host translation elongation initiation factors *eif4e* and *eif4g* are also recessive viral-resistance genes. Some have been deployed to control potyviruses in barley, rice, tomato, pepper, pea, lettuce and melon. The discovery prompted a successful mutant screen for chemically induced *eif4e* alleles in tomato.

Natural promoter variation can lead to the evolution of recessive disease-resistance alleles. For example, the recessive resistance gene *xa13* in rice is an allele of *Os-8N3*. *Os-8N3* is transcriptionally activated by*Xanthomonas oryzae pv. oryzae* strains that express the TAL effector *PthXo1*. The *xa13* gene has a mutated effector-binding element in its promoter that eliminates *PthXo1* binding and renders these lines resistant to strains that rely on *PthXo1*. This finding also demonstrated that *Os-8N3* is required for susceptibility.

Xa13/Os-8N3 is required for pollen development, showing that such mutant alleles can be problematic should the disease-susceptibility phenotype alter function in other processes. However, mutations in the *Os11N3* (OsSWEET14) TAL effector–binding element were made by fusing TAL effectors to nucleases (TALENs). Genome-edited rice plants with altered *Os11N3* binding sites remained resistant to *Xanthomonas oryzae pv. oryzae*, but still provided normal development function.

Gene Silencing

RNA silencing-based resistance is a powerful tool for engineering resistant crops. The advantage of RNAi as a novel gene therapy against fungal, viral and bacterial infection in plants lies in the fact that it regulates gene expression via messenger RNA degradation, translation repression and chromatin remodelling through small non-coding RNAs. Mechanistically, the silencing processes are guided by processing products of the double-stranded RNA (dsRNA) trigger, which are known as small interfering RNAs and microRNAs.

Host Range

Among the thousands of species of plant pathogenic microorganisms, only a small minority have the capacity to infect a broad range of plant species. Most pathogens instead exhibit a high degree of host-specificity. Non-host plant species are often said to express *non-host resistance*. The term *host resistance* is used when a pathogen species can be pathogenic on the host species but certain strains of that plant species resist certain strains of the pathogen species. The causes of host resistance and non-host resistance can overlap. Pathogen host range can change quite suddenly if, for example, the pathogen's capacity to synthesize a host-specific toxin or effector is gained by gene shuffling/mutation, or by horizontal gene transfer.

Epidemics and Population Biology

Native populations are often characterized by substantial genotype diversity and dispersed populations (growth in a mixture with many other plant species). They also have undergone of plant-pathogen coevolution. Hence as long as novel pathogens are not introduced/do not evolve, such populations generally exhibit only a low incidence of severe disease epidemics.

Monocrop agricultural systems provide an ideal environment for pathogen evolution, because they offer a high density of target specimens with similar/identical genotypes.

The rise in mobility stemming from modern transportation systems provides pathogens with access to more potential targets.

Climate change can alter the viable geographic range of pathogen species and cause some diseases to become a problem in areas where the disease was previously less important.

These factors make modern agriculture more prone to disease epidemics. Common solutions include constant breeding for disease resistance, use of pesticides, use of border inspections and plant import restrictions, maintenance of significant genetic diversity within the crop gene pool, and constant surveillance to accelerate initiation of appropriate responses. Some pathogen species have much greater capacity to overcome plant disease resistance than others, often because of their ability to evolve rapidly and to disperse broadly.

Disease Resistance in Fruit and Vegetables

There are a number of lines of defence against pests (that, those animals that cause damage to the plants we grow) and diseases in the O, principal among these being the practice of good husbandry, creating healthy soil and ensuring high standards of garden hygiene. But no matter how diverse and healthy the garden eco-system may be, there will always be a degree of disease and pest presence. In many ways, some level of pathogen population in the garden can be not only acceptable but desirable as they are indicative of a generally healthful and diverse environment, and add to the overall robustness of the system as an immunity to such detrimental influences will build up, particularly in a balanced polycultural regime. Indeed, most of the plants we grow will tend to be selected because they are trouble free, and those that are more susceptible to attack will have fallen by the wayside over time. However, most farmers find it unacceptable that the food crops they grow are damaged by pests.

For these crops there has been considerable research and selective breeding carried out in order to find cultivars that are resistant or immune to pest and disease damage. Breeding for plant disease resistance generally has involved finding suitable genetic material amongst existing stocks or in the wild, which is then incorporated into commercial varieties.

Example: The Apple

In the case of apples, in which research is being carried out in order to develop resistance to diseases such as apple scab (*Venturia inaequalis*), powdery mildew (*Podosphaera leucotricha*), orchard fireblight (*Erwinia amylovora*), woolly apple aphid (*Eriosoma lanigerum*) and collar rot (*Phytophthora cactorum*), the main sources of resistant material used in breeding programmes such as those being run by East Malling Research Station in England or Hortresearch in New Zealand are major gene resistances derived from crab-apples. The Vf gene for black spot resistance is derived from the ornamental crab-apple species *Malus floribunda*. Most black spot resistant cultivars developed around the world carry this gene, but there are some selections that carry the Vr (from M. pumila) or Vm (from M. micromalus) gene. Major gene resistances to powdery mildew are derived from M. robusta (Pl1) and M. zumi (Pl2), and the apple cultivar *Northern Spy* has a long-standing reputation for its major gene resistance to woolly apple aphid. Since early this century this resistance has been used to develop woolly aphid resistant rootstocks such as MM.106 and M.793. Much later it was shown that the cultivar was also very resistant to collar rot and a useful breeding parent for this resistance.

Resistance and Immunity

Some plants can tolerate the presence of large numbers of insects without being severely affected. This is not very satisfactory however as insects will still cause damage, and in fact further breeding and population expansion of the pest species is supported. Other varieties are less attractive to pests, but this can be difficult to sustain or demonstrate.The most valuable form of resistance is where the pest cannot survive as well on one variety as on another. In some cases this can actually make the plants immune to attack, as is the case with the lettuces *Avoncrisp* and *Avondefiance* which were bred at the Institute of Horticultural Research, Wellesbourne during the 1960s, which are fully resistant to lettuce root aphid (*Pemphigus bursarius*).

Trade-off of Breeding for Resistance

Sometimes however there can be a *trade-off*, for those varieties which have increased immunity or resistance may be lacking in other qualities such as flavour, yield or quality. Celery resistant to the Fusarium fungus (*Fusarium oxysporum spp.*) may not succumb to this disease, but may also be unacceptably short, ribby and low yielding.Further, a cultivar that is resistant to one disease may be more susceptible to another that is equally important. A lettuce cultivar that is resistant to mosaic virus may be sensitive to corky root disease, whilst another that resists corky root may be vulnerable to downy mildew (*Brim lactic*).Another drawback to resistance is that depending on the host pathogen system, resistance is sometimes not long lasting as new pathogen strains quickly develop, and further research and breeding is constantly needed.

Availability of Resistant Varieties

Resistant varieties are not available for all crops. For several of the most damaging plant diseases, such as Potato blight (*Phytophthora infestans*) and white rot (*Sclerotic cepivorum*) of the Allium family, no acceptable resistant cultivars are yet available. In addition, commercial seed companies and plant breeders rarely invest resources into developing resistant cultivars for more minor or speciality crops, which often tend to be those of greater interest to the organic grower.

In general it is probably fair to say that resistance will not fully guarantee total crop protection, but choosing resistant varieties should rather be considered as a part of an overall Integrated pest management strategy, especially against virus diseases. In particular they can be especially useful where the threat from specific pests and diseases is high.

Table of Resistant Crops

The table below illustrates a number of pest and disease resistant vegetable crop varieties, although should not be considered comprehensive.

Crop	Variety	Comments
French Bean	Aigullion	Resistant to bean mosaic virus and anthracnose
	Hildora	Tolerant of bean mosaic virus
	Maxi	Tolerant of bean mosaic virus
	Nektar Queen	Resistant to bean mosaic virus
	Farba	Resistant to bean mosaic virus
	Hilda	Resistant to bean mosaic virus
Broad Bean	Futura	Tolerant of chocolate spot
Brussels sprouts	Cavalier	Resistant to light leaf spot
	Braveheart F1	Resistant to powdery mildew and light leaf spot
Cabbage	Stonehead F1	Resistant to mildew
Calabrese	Emperor F1	Highly tolerant of black rot and downy mildew
	Sampson F1	Highly resistant to black rot and downy mildew
Carrot	Fly Away F1	Bred for resistance to carrot fly
	Newmarket	Good resistance to splitting
Cucumber	Bush Champion F1	Resistant to cucumber mosaic virus
	Marketmore	Resistant to cucumber mosaic virus
	Burpless Tasty Green F1	Resistant to mildew and tolerant of heat
	Slice King F1	Resistant to downy and powdery mildew, gummosis and angular leaf spot
	Cumlaude F1	Tolerant of powdery mildew
	Media F1	Tolerant of powdery mildew
Leek	Swiss Giant, Zermatt	Moderate rust resistance
	Alvitta	Rust Tolerant
	Conora	Good resistance to rust
Lettuce	Alexandria	Highly resistant to downy mildew and bolting
	Lisbusa	Resistant to downy mildew, tolerant of mosaic virus, root aphid and tipburn
	Sylvestra	Resistant to downy mildew and aphids
	Little Gem	Resistant to root aphid
	Barcelona	Resistant to aphids and mildew
	Iglo	Resistant to downey mildew
	Revolution	Resistant to mildew and bolting
	Smile	Resistant to aphids, mildew, tipburn and bolting
Parsnip	Avonreister	Very good resistance to canker
	Gladiator F1	High resistance to canker
	Tender and True	Good canker resistance
Pea	Ambassador	Resistant to powdery mildew, entation virus and fusarium wilt
	Cavalier	Good resistance to powdery mildew.
	Greenshaft	Resistant to downy mildew and fusarium wilt
	Rondo	Resistant to fusarium wilt
Potato (first early)	Colleen	Good resistance to blight and scab

	Premiere	Resistant to blight and spraing
	Pentland Javelin	Good all round disease resistance, including golden eelworm and scab
	Swift	Resistant to golden eelworm and tolerant to blackleg
Potato (second early)	Cosmos	Resistant to blight and common scab
	Osprey	Very resistant to scab and eelworm
Potato (maincrop)	Harmony	Very resistant to scab and partially resistant to white and golden eelworm
	Milva	Blight resistant
	Remarka	Good all around disease resistance
	Admiral	Good resistance to blight and scab
	Cara	Blight resistant
	Valor	Good overall disease resistance combined with high eelworm and blight resistance
Pepper	Bell Boy F1	Mosaic virus resistant
	Mandy F1	Highly resistant to mosaic virus
Sweetcorn	Golden sweet F1	Highly resistant to smut
Tomato	Alicante	Good resistance to greenback and mildew
	Libra F1	Resistant to fusarium crown and root rot
	Shirley F1	Resistant to fusarium, cladosporium and TMV

Physiological Plant Disorder

Deficit of micronutriens, vine.

Physiological plant disorders are caused by non-pathological conditions such as poor light, adverse weather, water-logging, phytotoxic compounds or a lack of nutrients, and affect the functioning of the plant system. Physiological disorders are distinguished from plant diseases caused by pathogens, such as a virus or fungus. While the symptoms of physiological disorders may appear disease-like, they can usually be prevented by altering environmental conditions. However, once a plant shows symptoms of a physiological disorder it is likely that that season's growth or yield will be reduced.

Diagnosis of the cause of a physiological disorder (or disease) can be difficult, but there are many web-based guides that may assist with this. Examples are: *Abiotic plant disorders: Symptoms, signs and solutions*; *Georgia Corn Diagnostic Guide*; *Diagnosing Plant Problems* (Kentucky); and *Diagnosing Plant Problems* (Virginia).

Sunburn at apple.

Some general tips to diagnosing plant disorders:

- Examine where symptoms first appear on a plant—on new leaves, old leaves or all over?
- Note the pattern of any discolouration or yellowing–is it all over, between the veins or around the edges? If only the veins are yellow deficiency is probably not involved.
- Note general patterns rather than looking at individual plants—are the symptoms distributed throughout a group of plants of the same type growing together. In the case of a deficiency all of the plants should be similarly effected, although distribution will depend on past treatments applied to the soil.
- Soil analysis, such as determining pH, can help to confirm the presence of physiological disorders.
- Consider recent conditions, such as heavy rains, dry spells, frosts, etc., may also help to determine the cause of plant disorders.

Weather Damage

Frost and cold are major causes of crop damage to tender plants, although hardy plants can also suffer if new growth is exposed to a hard frost following a period of warm weather. Symptoms will often appear overnight, affecting many types of plants. Leaves and stems may turn black, and buds and flowers may be discoloured, and frosted blooms may not produce fruit. Many annual plants, or plants grown in frost free areas, can suffer from damage when the air temperature drops below 40 degrees Fahrenheit (4 degrees Celsius). Tropical plants may begin to experience cold damage when the temperature is 42 to 48 °F (5 to 9 °C), symptoms include wilting of the top of the stems and/or leaves, and blackening or softening of the plant tissue.

Frost or cold damage can be avoided by ensuring that tender plants are properly hardened before planting, and that they are not planted too early in the season, before the risk of frost has passed. Avoid planting susceptible plants in frost pockets, or where they will receive early morning sun. Protect young buds and bloom with horticultural fleece if frost is forecast. Cold, drying easterly winds can also severely inhibit spring growth even without an actual frost, thus adequate shelter or the use of windbreaks is important.

Drought can cause plants to suffer from water stress and wilt. Adequate irrigation is required during prolonged hot, dry periods. Rather than shallow daily watering, during a drought water should be directed towards the roots, ensuring that the soil is thoroughly soaked two or three times a week. Mulches also help preserve soil moisture and keep roots cool.

Heavy rains, particularly after prolonged dry periods, can also cause roots to split, onion saddle-back (splitting at the base), tomatoes split and potatoes to become deformed or hollow. Using mulches or adding organic matter such as leaf mold, compost or well rotted manure to the soil will help to act as a 'buffer' between sudden changes in conditions. Water-logging can occur on poorly drained soils, particularly following heavy rains. Plants can become yellow and stunted, and will tend to be more prone to drought and diseases. Improving drainage will help to alleviate this problem.

Hail can cause damage to soft skinned fruits, and may also allow brown rot or other fungi to penetrate the plant. Brown spot markings or lines on one side of a mature apple are indicative of a spring hailstorm.

Plants affected by salt stress are unable to take water from soil, due to an osmotic imbalance between soil and plant.

Iron deficiency.

Nutrient Deficiencies

Poor growth and a variety of disorders such as leaf discolouration (chlorosis) can be caused by a shortage of one or more plant nutrients. Poor plant uptake of a nutrient from the soil (or other growing medium) may be due to an absolute shortage of that element in the growing medium, or because that element is present in a form that is not available to the plant. The latter can be caused by incorrect pH, shortage of water, poor root growth or an excess of another nutrient. Plant nutrient deficiencies can be avoided or corrected using a variety of approaches including the consultation of experts on-site, the use of soil and plant-tissue testing services, the application of prescription-blend fertilizers, the application of fresh or well-decomposed organic matter, and the use of biological systems such as cover crops, intercropping, improved fallows, ley cropping, permaculture, or crop rotation.

Drought.

Nutrient (or mineral) deficiencies include:

- Boron deficiency
- Calcium deficiency
- Iron deficiency
- Magnesium deficiency
- Manganese deficiency
- Nitrogen deficiency
- Phosphorus deficiency
- Potassium deficiency

- Zinc deficiency
- Shortage of trace elements such as molybdenum can also cause disorders such as whiptail in cauliflower.

Plant Defense Against Herbivory

Plant defense against herbivory or host-plant resistance (HPR) describes a range of adaptations evolved by plants which improve their survival and reproduction by reducing the impact of herbivores. Plants can sense being touched, and they can use several strategies to defend against damage caused by herbivores. Many plants produce secondary metabolites, known as allelochemicals, that influence the behavior, growth, or survival of herbivores. These chemical defenses can act as repellents or toxins to herbivores, or reduce plant digestibility.

Foxgloves produce several deadly chemicals, namely cardiac and steroidal glycosides. Ingestion can cause nausea, vomiting, hallucinations, convulsions, or death.

Other defensive strategies used by plants include escaping or avoiding herbivores in any time and/or any place, for example by growing in a location where plants are not easily found or accessed by herbivores, or by changing seasonal growth patterns. Another approach diverts herbivores toward eating non-essential parts, or enhances the ability of a plant to recover from the damage caused by herbivory. Some plants encourage the presence of natural enemies of herbivores, which in turn protect the plant. Each type of defense can be either *constitutive* (always present in the plant), or *induced* (produced in reaction to damage or stress caused by herbivores).

Historically, insects have been the most significant herbivores, and the evolution of land plants is closely associated with the evolution of insects. While most plant defenses are directed against insects, other defenses have evolved that are aimed at vertebrate herbivores, such as birds and mammals. The study of plant defenses against herbivory is important, not only from an evolutionary view point, but also in the direct impact that these defenses have on agriculture, including human and livestock food sources; as beneficial 'biological control agents' in biological pest control programs; as well as in the search for plants of medical importance.

Evolution of Defensive Traits

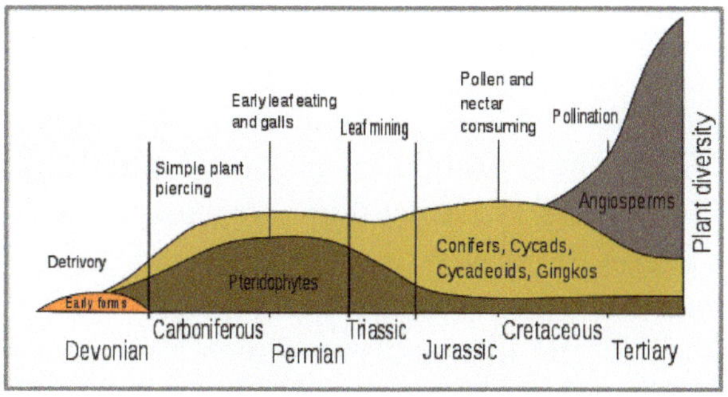

Timeline of plant evolution and the beginnings of different modes of insect herbivory

The earliest land plants evolved from aquatic plants around 450 million years ago (Ma) in the Ordovician period. Many plants have adapted to iodine-deficient terrestrial environment by removing iodine from their metabolism, in fact iodine is essential only for animal cells. An important antiparasitic action is caused by the block of the transport of iodide of animal cells inhibiting sodium-iodide symporter (NIS). Many plant pesticides are glycosides (as the cardiac digitoxin) and cyanogenic glycosides which liberate cyanide, which, blocking cytochrome c oxidase and NIS, is poisonous only for a large part of parasites and herbivores and not for the plant cells in which it seems useful in seed dormancy phase. Iodide is not pesticide, but is oxidized, by vegetable peroxidase, to iodine, which is a strong oxidant, it is able to kill bacteria, fungi and protozoa.

The Cretaceous period saw the appearance of more plant defense mechanisms. The diversification of flowering plants (angiosperms) at that time is associated with the sudden burst of speciation in insects. This diversification of insects represented a major selective force in plant evolution, and led to selection of plants that had defensive adaptations. Early insect herbivores were mandibulate and bit or chewed vegetation; but the evolution of vascular plants lead to the co-evolution of other forms of herbivory, such as sap-sucking, leaf mining, gall forming and nectar-feeding.

The relative abundance of different species of plants in ecological communities including forests and grasslands may be determined in part by the level of defensive compounds in the different species. Since the cost of replacement of damaged leaves is higher in conditions where resources are scarce, it may also be that plants growing in areas where water and nutrients are scarce may invest more resources into anti-herbivore defenses.

Records of Herbivores

Viburnum lesquereuxii leaf with insect damage; Dakota Sandstone (Cretaceous) of Ellsworth County, Kansas. Scale bar is 10 mm.

Our understanding of herbivory in geological time comes from three sources: fossilized plants, which may preserve evidence of defense (such as spines), or herbivory-related damage; the observation of plant debris in fossilised animal faeces; and the construction of herbivore mouthparts.

Long thought to be a Mesozoic phenomenon, evidence for herbivory is found almost as soon as fossils which could show it. Within under 20 million years of the first fossils of sporangia and stems towards the close of the Silurian, around 420 million years ago, there is evidence that they were being consumed. Animals fed on the spores of early Devonian plants, and the Rhynie chert also provides evidence that organisms fed on plants using a "pierce and suck" technique. Many plants of this time are preserved with spine-like enations, which may have performed a defensive role before being co-opted to develop into leaves.

During the ensuing 75 million years, plants evolved a range of more complex organs - from roots to seeds. There was a gap of 50 to 100 million years between each organ evolving, and it being fed upon. Hole feeding and skeletonization are recorded in the early Permian, with surface fluid feeding evolving by the end of that period.

A Plain Tiger *Danaus chrysippus* caterpillar making a moat to block defensive chemicals of *Calotropis* before feeding

Co-evolution

Herbivores are dependent on plants for food, and have evolved mechanisms to obtain this food despite the evolution of a diverse arsenal of plant defenses. Herbivore adaptations to plant defense have been likened to *offensive traits* and consist of adaptations that allow increased feeding and use of a host plant. Relationships between herbivores and their host plants often results in reciprocal evolutionary change, called co-evolution. When an herbivore eats a plant it selects for plants that can mount a defensive response. In cases where this relationship demonstrates *specificity* (the evolution of each trait is due to the other), and *reciprocity* (both traits must evolve), the species are thought to have co-evolved. The "escape and radiation" mechanism for co-evolution presents the idea that adaptations in herbivores and their host plants have been the driving force behind speciation, and have played a role in the radiation of insect species during the age of angiosperms. Some herbivores have evolved ways to hijack plant defenses to their own benefit, by sequestering these chemicals and using them to protect themselves from predators. Plant defenses against herbivores are generally not complete so plants also tend to evolve some tolerance to herbivory.

Types

Plant defenses can be classified generally as constitutive or induced. Constitutive defenses are always present in the plant, while induced defenses are produced or mobilized to the site where a plant is injured. There is wide variation in the composition and concentration of constitutive defenses and these range from mechanical defenses to digestibility reducers and toxins. Many external mechanical defenses and large quantitative defenses are constitutive, as they require large amounts of resources to produce and are difficult to mobilize. A variety of molecular and biochemical approaches are used to determine the mechanism of constitutive and induced plant defenses responses against herbivory.

Induced defenses include secondary metabolic products, as well as morphological and physiological changes. An advantage of inducible, as opposed to constitutive defenses, is that they are only produced when needed, and are therefore potentially less costly, especially when herbivory is variable.

Chemical Defenses

The evolution of chemical defences in plants is linked to the emergence of chemical substances that are not involved in the essential photosynthetic and metabolic activities. These substances, secondary metabolites, are organic compounds that are not directly involved in the normal growth, development or reproduction of organisms, and often produced as by-products during the synthesis of primary metabolic products. Although these secondary metabolites have been thought to play a major role in defenses against herbivores, a meta-analysis of recent relevant studies has suggested that they have either a more minimal (when compared to other non-secondary metabolites, such as primary chemistry and physiology) or more complex involvement in defense.

Persimmon, genus *Diospyros*, has a high tannin content which gives immature fruit, seen above, an astringent and bitter flavor.

Secondary metabolites are often characterized as either *qualitative* or *quantitative*. Qualitative metabolites are defined as toxins that interfere with an herbivore's metabolism, often by blocking specific biochemical reactions. Qualitative chemicals are present in plants in relatively low concentrations (often less than 2% dry weight), and are not dosage dependent. They are usually small, water-soluble molecules, and therefore can be rapidly synthesized, transported and stored with relatively little energy cost to the plant. Qualitative allelochemicals are usually effective against non-adapted specialists and generalist herbivores.

Quantitative chemicals are those that are present in high concentration in plants (5 – 40% dry weight) and are equally effective against all specialists and generalist herbivores. Most quantitative metabolites are digestibility reducers that make plant cell walls indigestible to animals. The effects of quantitative metabolites are dosage dependent and the higher these chemicals' proportion in the herbivore's diet, the less nutrition the herbivore can gain from ingesting plant tissues. Because they are typically large molecules, these defenses are energetically expensive to produce and maintain, and often take longer to synthesize and transport.

The geranium, for example, produces a unique chemical compound in its petals to defend itself from Japanese beetles. Within 30 minutes of ingestion the chemical paralyzes the herbivore. While the chemical usually wears off within a few hours, during this time the beetle is often consumed by its own predators.

Types of Chemical Defences

Plants have evolved many secondary metabolites involved in plant defense, which are collectively known as antiherbivory compounds and can be classified into three sub-groups: nitrogen compounds (including *alkaloids*, *cyanogenic glycosides*, *glucosinolates* and *benzoxazinoids*), terpenoids, and phenolics.

Alkaloids are derived from various amino acids. Over 3000 known alkaloids exist, examples include nicotine, caffeine, morphine, cocaine, colchicine, ergolines, strychnine, and quinine. Alkaloids have pharmacological effects on humans and other animals. Some alkaloids can inhibit or activate enzymes, or alter carbohydrate and fat storage by inhibiting the formation phosphodiester bonds involved in their breakdown. Certain alkaloids bind to nucleic acids and can inhibit synthesis of proteins and affect DNA repair mechanisms. Alkaloids can also affect cell membrane and cytoskeletal structure causing the cells to weaken, collapse, or leak, and can affect nerve transmission. Although alkaloids act on a diversity of metabolic systems in humans and other animals, they almost uniformly invoke an aversively bitter taste.

Cyanogenic glycosides are stored in inactive forms in plant vacuoles. They become toxic when herbivores eat the plant and break cell membranes allowing the glycosides to come into contact with enzymes in the cytoplasm releasing hydrogen cyanide which blocks cellular respiration. Glucosinolates are activated in much the same way as cyanogenic glucosides, and the products can cause gastroenteritis, salivation, diarrhea, and irritation of the mouth. Benzoxazinoids, secondary defence metabolites, which are characteristic for grasses (Poaceae), are also stored as inactive glucosides in the plant vacuole. Upon tissue disruption they get into contact with β-glucosidases from the chloroplasts, which enzymatically release the toxic aglucones. Whereas some benzoxazinoids are constitutively present, others are only synthesised following herbivore infestation, and thus, considered inducible plant defenses against herbivory.

The terpenoids, sometimes referred to as isoprenoids, are organic chemicals similar to terpenes, derived from five-carbon isoprene units. There are over 10,000 known types of terpenoids. Most are multicyclic structures which differ from one another in both functional groups, and in basic carbon skeletons. Monoterpenoids, continuing 2 isoprene units, are volatile essential oils such as citronella, limonene, menthol, camphor, and pinene. Diterpenoids, 4 isoprene units, are widely distributed in latex and resins, and can be quite toxic. Diterpenes are responsible for making *Rhododendron* leaves poisonous. Plant steroids and sterols are also produced from terpenoid precursors, including vitamin D, glycosides (such as digitalis) and saponins (which lyse red blood cells of herbivores).

Phenolics, sometimes called *phenols*, consist of an aromatic 6-carbon ring bonded to a hydroxy group. Some phenols have antiseptic properties, while others disrupt endocrine activity. Phenolics range from simple tannins to the more complex flavonoids that give plants much of their red, blue, yellow, and white pigments. Complex phenolics called polyphenols are capable of producing many different types of effects on humans, including antioxidant properties. Some examples of phenolics used for defense in plants are: lignin, silymarin and cannabinoids. Condensed tannins, polymers composed of 2 to 50 (or more) flavonoid molecules, inhibit herbivore digestion by binding to consumed plant proteins and making them more difficult for animals to digest, and by interfering with protein absorption and digestive enzymes. Silica and lignins, which are completely indigestible to animals, grind down insect mandibles (appendages necessary for feeding).

In addition to the three larger groups of substances mentioned above, fatty acid derivates, amino acids and even peptides are also used as defense. The cholinergic toxine, cicutoxin of water hemlock, is a polyyne derived from the fatty acid metabolism. diaminopropionic acid as simple amino acid is used by the sweet pea which leads also to intoxication in humans. The synthesis of fluoroacetate in several plants is an example of the use of small molecules to disrupt the metabolism of herbivores, in this case the citric acid cycle.

In tropical *Sargassum* and *Turbinaria* species that are often preferentially consumed by herbivorous fishes and echinoids, there is a relatively low level of phenolics and tannins.

Mechanical Defenses

The thorns on the stem of this raspberry plant, serve as a mechanical defense against herbivory.

Plants have many external structural defenses that discourage herbivory. Depending on the herbivore's physical characteristics (i.e. size and defensive armor), plant structural defenses on stems and leaves can deter, injure, or kill the grazer. Some defensive compounds are produced internally but are released onto the plant's surface; for example, resins, lignins, silica, and wax cover the epidermis of terrestrial plants and alter the texture of the plant tissue. The leaves of holly plants, for instance, are very smooth and slippery making feeding difficult. Some plants produce gummosis or sap that traps insects.

A plant's leaves and stem may be covered with sharp prickles, spines, thorns, or trichomes- hairs on the leaf often with barbs, sometimes containing irritants or poisons. Plant structural features like spines and thorns reduce feeding by large ungulate herbivores (e.g. kudu, impala, and goats) by restricting the herbivores' feeding rate, or by wearing down the molars. Raphides are sharp needles of calcium oxalate or calcium carbonate in plant tissues, making ingestion painful, damaging a herbivore's mouth and gullet and causing more efficient delivery of the plant's toxins. The structure of a plant, its branching and leaf arrangement may also be evolved to reduce herbivore impact. The shrubs of New Zealand have evolved special wide branching adaptations believed to be a response to browsing birds such as the moas. Similarly, African Acacias have long spines low in the canopy, but very short spines high in the canopy, which is comparatively safe from herbivores such as giraffes.

Coconut palms protect their fruit by surrounding it with multiple layers of armor.

Trees such as coconut and other palms, may protect their fruit by multiple layers of armor, needing efficient tools to break through to the seed contents, and special skills to climb the tall and relatively smooth trunk.

Thigmonasty

Thigmonastic movements, those that occur in response to touch, are used as a defense in some plants. The leaves of the sensitive plant, *Mimosa pudica*, close up rapidly in response to direct touch, vibration, or even electrical and thermal stimuli. The proximate cause of this mechanical response is an abrupt change in the turgor pressure in the pulvini at the base of leaves resulting from osmotic phenomena. This is then spread via both electrical and chemical means through the plant; only a single leaflet need be disturbed.

This response lowers the surface area available to herbivores, which are presented with the underside of each leaflet, and results in a wilted appearance. It may also physically dislodge small herbivores, such as insects.

Mimicry and Camouflage

Some plants mimic the presence of insect eggs on their leaves, dissuading insect species from laying their eggs there. Because female butterflies are less likely to lay their eggs on plants that already have butterfly eggs, some species of neotropical vines of the genus *Passiflora* (Passion flowers) contain physical structures resembling the yellow eggs of *Heliconius* butterflies on their leaves, which discourage oviposition by butterflies.

Indirect Defenses

The large thorn-like stipules of *Acacia collinsii* are hollow and offer shelter for ants, which in return protect the plant against herbivores.

Another category of plant defenses are those features that indirectly protect the plant by enhancing the probability of attracting the natural enemies of herbivores. Such an arrangement is known as mutualism, in this case of the "enemy of my enemy" variety. One such feature are semiochemicals, given off by plants. Semiochemicals are a group of volatile organic compounds involved in interactions between organisms. One group of semiochemicals are allelochemicals; consisting of allomones, which play a defensive role in interspecies communication, and kairomones, which are used by members of higher trophic levels to locate food sources. When a plant is attacked it releases allelochemics containing an abnormal ratio of these herbivore-induced plant volatiles (HIPVs). Predators sense these volatiles as food cues, attracting them to the damaged plant, and to feeding herbivores. The subsequent reduction in the number of herbivores confers a fitness benefit to the plant and demonstrates the indirect defensive capabilities of semiochemicals. Induced volatiles also have drawbacks, however; some studies have suggested that these volatiles also attract herbivores.

Plants also provide housing and food items for natural enemies of herbivores, known as "biotic" defense mechanisms, as a means to maintain their presence. For example, trees from the genus *Macaranga* have adapted their thin stem walls to create ideal housing for an ant species (genus *Crematogaster*), which, in turn, protects the plant from herbivores. In addition to providing housing, the plant also provides the ant with its exclusive food source; from the food bodies produced by the plant. Similarly, some *Acacia* tree species have developed thorns that are swollen at the base, forming a hollowing structure that acts as housing. These *Acacia* trees also produce nectar in extrafloral nectaries on their leaves as food for the ants.

Plant use of endophytic fungi in defense is a very common phenomenon. Most plants have endophytes, microbial organisms that live within them. While some cause disease, others protect plants from herbivores and pathogenic microbes. Endophytes can help the plant by producing toxins harmful to other organisms that would attack the plant, such as alkaloid producing fungi which are common in grasses such as tall fescue (*Festuca arundinacea*).

Leaf Shedding and Color

There have been suggestions that leaf shedding may be a response that provides protection against diseases and certain kinds of pests such as leaf miners and gall forming insects. Other responses such as the change of leaf colors prior to fall have also been suggested as adaptations that may help undermine the camouflage of herbivores. Autumn leaf color has also been suggested to act as an honest warning signal of defensive commitment towards insect pests that migrate to the trees in autumn.

Costs and Benefits

Defensive structures and chemicals are costly as they require resources that could otherwise be used by plants to maximize growth and reproduction. Many models have been proposed to explore how and why some plants make this investment in defenses against herbivores.

Optimal Defense Hypothesis

The optimal defense hypothesis attempts to explain how the kinds of defenses a particular plant might use reflect the threats each individual plant faces. This model considers three main factors, namely: risk of attack, value of the plant part, and the cost of defense.

The first factor determining optimal defense is risk: how likely is it that a plant or certain plant parts will be attacked? This is also related to the *plant apparency hypothesis*, which states that a plant will invest heavily in broadly effective defenses when the plant is easily found by herbivores. Examples of apparent plants that produce generalized protections include long-living trees, shrubs, and perennial grasses. Unapparent plants, such as short-lived plants of early successional stages, on the other hand, preferentially invest in small amounts of qualitative toxins that are effective against all but the most specialized herbivores.

The second factor is the value of protection: would the plant be less able to survive and reproduce after removal of part of its structure by a herbivore? Not all plant parts are of equal evolutionary value, thus valuable parts contain more defenses. A plant's stage of development at the time of feeding also affects the resulting change in fitness. Experimentally, the fitness value of a plant structure is determined by removing that part of the plant and observing the effect. In general, reproductive parts are not as easily replaced as vegetative parts, terminal leaves have greater value than basal leaves, and the loss of plant parts mid-season has a greater negative effect on fitness than removal at the beginning or end of the season. Seeds in particular tend to be very well protected. For example, the seeds of many edible fruits and nuts contain cyanogenic glycosides such as amygdalin. This results from the need to balance the effort needed to make the fruit attractive to animal dispersers while ensuring that the seeds are not destroyed by the animal.

The final consideration is cost: how much will a particular defensive strategy cost a plant in energy and materials? This is particularly important, as energy spent on defense cannot be used for other functions, such as reproduction and growth. The optimal defense hypothesis predicts that plants will allocate more energy towards defense when the benefits of protection outweigh the costs, specifically in situations where there is high herbivore pressure.

Carbon:nutrient Balance Hypothesis

The carbon:nutrient balance hypothesis, also known as the *environmental constraint hypothesis* or *Carbon Nutrient Balance Model* (CNBM), states that the various types of plant defenses are responses to variations in the levels of nutrients in the environment. This hypothesis predicts the Carbon/Nitrogen ratio in plants determines which secondary metabolites will be synthesized. For example, plants growing in nitrogen-poor soils will use carbon-based defenses (mostly digestibility reducers), while those growing in low-carbon environments (such as shady conditions) are more likely to produce nitrogen-based toxins. The hypothesis further predicts that plants can change their defenses in response to changes in nutrients. For example, if plants are grown in low-nitrogen conditions, then these plants will implement a defensive strategy composed of constitutive carbon-based defenses. If nutrient levels subsequently increase, by for example the addition of fertilizers, these carbon-based defenses will decrease.

Growth Rate Hypothesis

The growth rate hypothesis, also known as the *resource availability hypothesis*, states that defense strategies are determined by the inherent growth rate of the plant, which is in turn determined by the resources available to the plant. A major assumption is that available resources are the limiting factor in determining the maximum growth rate of a plant species. This model predicts that the level of defense investment will increase as the potential of growth decreases. Additionally, plants in resource-poor areas, with inherently slow-growth rates, tend to have long-lived leaves and twigs, and the loss of plant appendages may result in a loss of scarce and valuable nutrients.

A recent test of this model involved a reciprocal transplants of seedlings of 20 species of trees between clay soils (nutrient rich) and white sand (nutrient poor) to determine whether trade-offs between growth rate and defenses restrict species to one habitat. When planted in white sand and protected from herbivores, seedlings originating from clay outgrew those originating from the nutrient-poor sand, but in the presence of herbivores the seedlings originating from white sand performed better, likely due to their higher levels of constitutive carbon-based defenses. These finding suggest that defensive strategies limit the habitats of some plants.

Growth-differentiation Balance Hypothesis

The growth-differentiation balance hypothesis states that plant defenses are a result of a tradeoff between "growth-related processes" and "differentiation-related processes" in different environments. Differentiation-related processes are defined as "processes that enhance the structure or function of existing cells (i.e. maturation and specialization)." A plant will produce chemical defenses only when energy is available from photosynthesis, and plants with the highest concentrations of secondary metabolites are the ones with an intermediate level of available resources. The GDBH also accounts for tradeoffs between growth and defense over a resource availability gradient. In situations where resources (e.g. water and nutrients) limit photosynthesis, carbon supply is predicted to limit both growth and defense. As resource availability increases, the requirements needed to support photosynthesis are met, allowing for accumulation of carbohydrate in tissues. As resources are not sufficient to meet the large demands of growth, these carbon compounds can instead be partitioned into the synthesis of carbon based secondary metabolites (phenolics, tannins, etc.). In environments where the resource demands for growth are met, carbon is allocated to rapidly dividing meristems (high sink strength) at the expense of secondary metabolism. Thus rapidly growing plants are predicted to contain lower levels of secondary metabolites and vice versa. In addition, the tradeoff predicted by the GDBH may change over time, as evidenced by a recent study on Salix spp. Much support for this hypothesis is present in the literature, and some scientists consider the GDBH the most mature of the plant defense hypotheses.

Importance to Humans

Agriculture

The variation of plant susceptibility to pests was probably known even in the early stages of agriculture in humans. In historic times, the observation of such variations in susceptibility have provided solutions for major socio-economic problems. The grape phylloxera was introduced from North America to France in 1860 and in 25 years it destroyed nearly a third (100,000 km^2) of the French grape yards. Charles Valentine Riley noted that the American species *Vitis labrusca* was resistant to *Phylloxera*. Riley, with J. E. Planchon, helped save the French wine industry by suggesting the grafting of the susceptible but high quality grapes onto *Vitis labrusca* root stocks. The formal study of plant resistance to herbivory was first covered extensively in 1951 by Reginald (R.H.) Painter, who is widely regarded as the founder of this area of research, in his book *Plant Resistance to Insects*. While this work pioneered further research in the US, the work of Chesnokov was the basis of further research in the USSR.

Fresh growth of grass is sometimes high in prussic acid content and can cause poisoning of grazing livestock. The production of cyanogenic chemicals in grasses is primarily a defense against herbivores.

The human innovation of cooking may have been particularly helpful in overcoming many of the defensive chemicals of plants. Many enzyme inhibitors in cereal grains and pulses, such as trypsin inhibitors prevalent in pulse crops, are denatured by cooking, making them digestible.

It has been known since the late 17th century that plants contain noxious chemicals which are avoided by insects. These chemicals have been used by man as early insecticides; in 1690 nicotine was extracted from tobacco and used as a contact insecticide. In 1773, insect infested plants were treated with nicotine fumigation by heating tobacco and blowing the smoke over the plants.

The flowers of *Chrysanthemum* species contain pyrethrin which is a potent insecticide. In later years, the applications of plant resistance became an important area of research in agriculture and plant breeding, particularly because they can serve as a safe and low-cost alternative to the use of pesticides. The important role of secondary plant substances in plant defense was described in the late 1950s by Vincent Dethier and G.S. Fraenkel. The use of botanical pesticides is widespread and notable examples include Azadirachtin from the neem (*Azadirachta indica*), d-Limonene from Citrus species, Rotenone from *Derris*, Capsaicin from chili pepper and Pyrethrum.

Natural materials found in the environment also induce plant resistance as well. Chitosan derived from chitin induce a plant's natural defense response against pathogens, diseases and insects including cyst nematodes, both are approved as biopesticides by the EPA to reduce the dependence on toxic pesticides.

The selective breeding of crop plants often involves selection against the plant's intrinsic resistance strategies. This makes crop plant varieties particularly susceptible to pests unlike their wild relatives. In breeding for host-plant resistance, it is often the wild relatives that provide the source of resistance genes. These genes are incorporated using conventional approaches to plant breeding, but have also been augmented by recombinant techniques, which allow introduction of genes from completely unrelated organisms. The most famous transgenic approach is the introduction of genes from the bacterial species, *Bacillus thuringiensis*, into plants. The bacterium produces proteins that, when ingested, kill lepidopteran caterpillars. The gene encoding for these highly toxic proteins, when introduced into the host plant genome, confers resistance against caterpillars, when the same toxic proteins are produced within the plant. This approach is controversial, however, due to the possibility of ecological and toxicological side effects.

Pharmaceutical

Illustration from the 15th-century manuscript *Tacuinum Sanitatis* detailing the beneficial and harmful properties of Mandrakes

Many currently available pharmaceuticals are derived from the secondary metabolites plants use to protect themselves from herbivores, including opium, aspirin, cocaine, and atropine. These chemicals have evolved to affect the biochemistry of insects in very specific ways. However, many of these biochemical pathways are conserved in vertebrates, including humans, and the chemicals act on human biochemistry in ways similar to that of insects. It has therefore been suggested that the study of plant-insect interactions may help in bioprospecting.

There is evidence that humans began using plant alkaloids in medical preparations as early as 3000 B.C. Although the active components of most medicinal plants have been isolated only recently (beginning in the early 19th century) these substances have been used as drugs throughout the human history in potions, medicines, teas and as poisons. For example, to combat herbivory by the larvae of some Lepidoptera species, Cinchona trees produce a variety of alkaloids, the most familiar of which is quinine. Quinine is extremely bitter, making the bark of the tree quite unpalatable, it is also an anti-fever agent, known as Jesuit's bark, and is especially useful in treating malaria.

Throughout history mandrakes (*Mandragora officinarum*) have been highly sought after for their reputed aphrodisiac properties. However, the roots of the mandrake plant also contain large quantities of the alkaloid scopolamine, which, at high doses, acts as a central nervous system depressant, and makes the plant highly toxic to herbivores. Scopolamine was later found to be medicinally used for pain management prior to and during labor; in smaller doses it is used to prevent motion sickness. One of the most well-known medicinally valuable terpenes is an anticancer drug, taxol, isolated from the bark of the Pacific yew, *Taxus brevifolia*, in the early 1960s.

Biological Pest Control

Repellent companion planting, defensive live fencing hedges, and "obstructive-repellent" interplanting, with host-plant resistance species as beneficial 'biological control agents' is a technique in biological pest control programs for: organic gardening, wildlife gardening, sustainable gardening, and sustainable landscaping; in organic farming and sustainable agriculture; and in restoration ecology methods for habitat restoration projects.

References

- Alberts B, Johnson A, Lewis J, Raff M, Roberts K, Walter P (2002). Molecular Biology of the Cell. Garland Science. ISBN 0-8153-3218-1.
- Fred Baur. Insect Management for Food Storage and Processing. American Association of Cereal Chemists. ISBN 0-913250-38-4.
- Jackson RW (editor). (2009). Plant Pathogenic Bacteria: Genomics and Molecular Biology. Caister Academic Press. ISBN 978-1-904455-37-0.
- Madden, Laurence; Gareth Hughes; Frank Van Den Bosch (2007). Study of Plant Disease Epidemics. American Phytopathological Society. ISBN 978-0-89054-354-2.
- Mengel, K.; Kirkby, E.A. (2001). Principles of plant nutrition (5th ed.). Dordrecht: Kluwer Academic Publishers. ISBN 079237150X.
- Futuyma, Douglas J.; Montgomery Slatkin (1983). Coevolution. Sunderland, Massachusetts: Sinauer Associates. ISBN 0-87893-228-3.
- Whittaker, Robert H. (1970). "The biochemical ecology of higher plants". In Ernest Sondheimer; John B. Simeone. Chemical ecology. Boston: Academic Press. pp. 43–70. ISBN 0-12-654750-5.
- Roberts, Margaret F.; Michael Wink (1998). Alkaloids: biochemistry, ecology, and medicinal applications. New York: Plenum Press. ISBN 0-306-45465-3.

- Gershenzon, Jonathan; Wolfgang Kreis (1999). "Biochemistry of terpenoids". In Michael Wink. Biochemistry of plant secondary metabolism. London: Sheffield Academic Press. pp. 222–279. ISBN 0-8493-4085-3.
- Raven, Peter H.; Ray F. Evert; Susan E. Eichhorn (2005). Biology of Plants. New York: W. H. Freeman and Company. ISBN 0-7167-1007-2.
- D., McKey (1979). "The distribution of secondary compounds within plants.". In Gerald A. Rosenthal; Daniel H. Janzen. Herbivores, their interaction with secondary plant metabolites. Boston: Academic Press. pp. 55– 133. ISBN 0-12-597180-X.
- Krischik, V. A.; R. F. Denno (1983). "Individual, population, and geographic patterns in plant defense.". In Robert F. Denno; Mark S. McClure. Variable plants and herbivores in natural and managed systems. Boston: Academic Press. pp. 463–512. ISBN 0-12-209160-4.
- Michael Smith, C. (2005). Plant Resistance to Arthropods: Molecular and Conventional Approaches. Berlin: Springer. ISBN 1-4020-3701-5.
- "Fusarium oxysporum : The End of the Banana Industry?". The Origin of Plant Pathology and The Potato Famine, and Other Stories of Plant Diseases. Retrieved 4 February 2015.
- Lee, R.D. (2012). "Georgia Corn Diagnostic Guide". extension.uga.edu. University of Georgia Cooperative Extension. Retrieved 10 April 2015.
- Green, J.L.; Maloy, O.; Capizzi, J.; Hartman, J.; Townsend, L. (2011). "Diagnosing Plant Problems: Kentucky Master Gardener Manual Chapter 7" (PDF). www2.ca.uky.edu. Kentucky Cooperative Extension. Retrieved 10 April 2015.

Permissions

All chapters in this book are published with permission under the Creative Commons Attribution Share Alike License or equivalent. Every chapter published in this book has been scrutinized by our experts. Their significance has been extensively debated. The topics covered herein carry significant information for a comprehensive understanding. They may even be implemented as practical applications or may be referred to as a beginning point for further studies.

We would like to thank the editorial team for lending their expertise to make the book truly unique. They have played a crucial role in the development of this book. Without their invaluable contributions this book wouldn't have been possible. They have made vital efforts to compile up to date information on the varied aspects of this subject to make this book a valuable addition to the collection of many professionals and students.

This book was conceptualized with the vision of imparting up-to-date and integrated information in this field. To ensure the same, a matchless editorial board was set up. Every individual on the board went through rigorous rounds of assessment to prove their worth. After which they invested a large part of their time researching and compiling the most relevant data for our readers.

The editorial board has been involved in producing this book since its inception. They have spent rigorous hours researching and exploring the diverse topics which have resulted in the successful publishing of this book. They have passed on their knowledge of decades through this book. To expedite this challenging task, the publisher supported the team at every step. A small team of assistant editors was also appointed to further simplify the editing procedure and attain best results for the readers.

Apart from the editorial board, the designing team has also invested a significant amount of their time in understanding the subject and creating the most relevant covers. They scrutinized every image to scout for the most suitable representation of the subject and create an appropriate cover for the book.

The publishing team has been an ardent support to the editorial, designing and production team. Their endless efforts to recruit the best for this project, has resulted in the accomplishment of this book. They are a veteran in the field of academics and their pool of knowledge is as vast as their experience in printing. Their expertise and guidance has proved useful at every step. Their uncompromising quality standards have made this book an exceptional effort. Their encouragement from time to time has been an inspiration for everyone.

The publisher and the editorial board hope that this book will prove to be a valuable piece of knowledge for students, practitioners and scholars across the globe.

Index